D1275236

100 ALIEN INVADERS

GILL WILLIAMS

Bradt

First published August 2011
Bradt Travel Guides Ltd
1st Floor, IDC House, The Vale, Chalfont St Peter, Bucks SL9 9RZ, England
www.bradtguides.com
Published in the USA by The Globe Pequot Press Inc,
PO Box 480, Guilford, Connecticut 06437-0480

Most images in this book were kindly supplied by Frank Lane Picture Agency (www.flpa-images.co.uk).
For details of all photographers see page 158.

ISBN: 978 1 84162 359 7

British Library Cataloguing in Publication Data
A catalogue record for this book is available from the British Library

Front cover, main image: American bullfrog (Thomas Marent/Minden Pictures/FLPA)
Front cover, other images, clockwise from top left: grey squirrel (Andrew Parkinson/FLPA); barn owl (Paul Hobson/FLPA); Louisiana crayfish (Fabio Pupin/FLPA); Burmese python (ZSSD/Minden Pictures/FLPA)

Back cover: green iguana (the Natural History Museum/Alamy)

Designed and formatted by Chris Lane at Artinfusion
Maps by Chris Lane

Production managed by Jellyfish Print Solutions; printed in India

Foreword

The environmental impacts of the way in which people choose to live on Earth are accelerating. Of particular concern to all who care about a sustainable future for people is the unwanted movement of invasive animals and plants.

These are the species highlighted in this book – 100 of the most significant alien invaders on the planet. Together, they account for massive global change, and many command significant costs in attempts to control or eradicate them. The story of prickly pear in the country of my birth, Australia, and its sequel with the introduction of the cane toad, are both notable examples described here.

Animals and plants removed from their natural predators and pathogens can run riot, causing enormous human misery, changing habitats and nutrient flows, and devastating ever more threatened native biodiversity. We need to be aware of such threats and manage them sensibly. Understanding what has happened in the past is an important step towards enlightened management of tomorrow's new alien invaders.

At Kew, together with our international partners, we are marshalling our systems and collections to better provide access to information about the world's plants and fungi. This includes their geographical distributions, conservation, restoration and sustainable use, as well as their potential – realised or latent – to become alien invaders. We actively manage weeds and invasive pests at our two World Heritage botanic gardens, as well as with partners in many countries and overseas territories. Our research aims to make a difference in the management and control of alien invaders, and we therefore fully support the sentiments behind *100 Alien Invaders*.

I commend this handsomely produced book, and congratulate all involved in its publication. It will, I'm sure, stimulate awareness and appropriate action in a compelling way on one of the most important biological issues of our time.

Steve Hopper

Professor Stephen D Hopper
Director, Royal Botanic Gardens, Kew

Contents

Foreward 3
Introducing the aliens 6

1 INVADERS IN COLD BLOOD 8

Introduction to reptiles & amphibians 10
Alien invaders 1–9 12
Protecting the Galápagos 24
Alien invaders 10–11 26

2 FURRY FIENDS AND FELONS 28

Introduction to mammals 30
Alien invaders 12–37 32

3 UNDERWATER ALIENS 60

Introduction to underwater aliens 62
Alien invaders 38–45 64
Coral reefs under threat 72
Alien invaders 46–49 74
The explorers 78

4 MARCH OF THE CREEPY-CRAWLIES 80

Introduction to creepy-crawlies 82
Alien invaders 50–62 84
Saving the Everglades 96

5 OUR FEATHERED FOES 98

Introduction to birds 100
Alien invaders 63–73 102
South Atlantic 114
Alien invaders 74–78 116
The Victorians 122

6 GATE-CRASHING GREENERY 124

Introduction to plants 126
Alien invaders 79–80 128
Kew Gardens 130
Alien invaders 81–85 132
Gardening for biodiversity 136
Alien invaders 86–91 138
Antarctica: time to close the freezer door 144
Alien invaders 92–100 146

Conservation contacts 155
Acknowledgements and picture credits 158
Index 159

An eye for a bird: the stoat has taken a heavy toll of New Zealand's kiwis since it was introduced in the 1880s.

INTRODUCING THE ALIENS

"It is not the strongest of the species that survives, nor the most intelligent that survives. It is the one that is the most adaptable to change."
Charles Darwin, *On The Origin Of Species*

Invasive species are dangerous aliens. They form one of the three greatest threats to life on Earth, right up there with climate change and the mushrooming growth of our human population.

Force of nature

So what is an invasive species? In simple terms it is an introduced plant or animal that causes devastating environmental problems in its adopted home. The lethal spread of these unwelcome organisms has all the hallmarks of a horror B movie. Alien invaders are able to topple entire ecosystems by upsetting the balance of nature. This might be by gobbling up a vital link in the food chain in a lake, or altering the soil composition of an ancient forest. An invasive animal might be smaller than a pinhead, yet able to spread killer diseases that can wipe out many species or devastate whole economies. These aliens outcompete or prey on vulnerable native species, and their impact goes on year after year, escalating as the invaders spread their net.

Invasive species have brought about the extinction of thousands of plants and animals right across the globe during the last 500 years. In the 21st century, pressure from exotic species is the biggest threat to the survival of nearly half the world's most endangered plants and animals.

▲ *A nose for success: the brown rat has expanded its worldwide range as a stowaway on ships, thriving on our rubbish and leftovers.*

Generalists versus specialists

The plants and animals poised to take over the Earth are those that are the most flexible and able to cope with rapid change. This adaptability enables them to colonise new parts of the world, and to threaten the survival of native species that are more set in their ways.

Most invaders are generalists – they can live in just about any climate, from the tropics to polar regions, and eat all sorts of food. The big losers tend to be specialist species, limited to very particular habitats and narrow diets. Specialists face a struggle for survival when generalists arrive and compete for the same food and shelter.

Aliens in numbers

50 The average number of threatening invasive species in each country worldwide.

10,000 The number of invasive species in Europe, of which 1,300 are damaging the environment, local economies or health.

75% The percentage rise in Europe's alien species since 1980.

Most generalists come from large land areas, where they've evolved to deal with lots of competition and many predators. So the biggest impacts of invasive species are felt on smaller islands. Here, the specialist natives haven't needed to develop systems to cope with tough competition. Instead, they've evolved to rub along nicely with just a few familiar species in a small environment. They're not able or equipped to move with the times.

So who's to blame?

People, of course. For millions of years, the spread of plants and animals was limited by oceans, deserts and mountain ranges. Only a handful of species were able to cross these natural barriers. Then humans emerged. By the Middle Ages, we were traversing oceans and introducing plants and animals to isolated ecosystems right across the Earth. Today, species are whizzing around the world faster than ever before, thanks to modern transport and the globalisation of trade.

Among the most successful aliens are those that live around people. Some, like rats and mice, are unwelcome stowaways, but many have been introduced deliberately, for farms or the pet trade. In the first six years of the 21st century, the United States alone imported 1.5 billion live animals. Several of the world's most damaging invaders have been released into new countries to deal with crop bugs, only to become a much worse pest. One such notorious mistake was the transporting of the unlovable Cane toad into Australia (see page 14).

How to use the maps

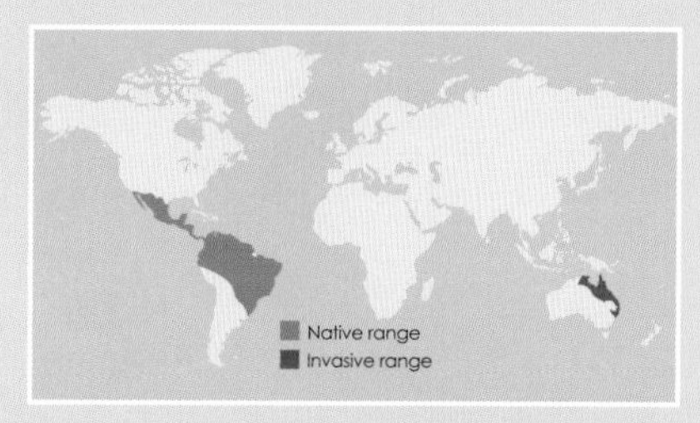

A distribution map is provided for each of the 100 species in this book, with the green area marking its native range and the red area its invasive range. These maps aim to give a broad-brush idea of where each species originated and where it has since spread, while highlighting a few key locations mentioned in the text. However, data is not available for all species in all regions, and thus the maps cannot be entirely accurate at a country level. Where the whole of a particular country is coloured red or green, this means that the species is known to be present but not necessarily in all parts. The mapping of invasive aliens is a dynamic science and many species have the potential to spread much further than the range indicated here.

▲ *Pretty Polly? The ring-necked parakeet may look attractive but its invasion of UK gardens means problems for native birds.*

Hitting our pockets

Alien invaders are not just dangerous for the environment. They also do enormous economic damage – about US $1.4 trillion every year, or 5% of the whole world's economy. Invasive species destroy crops and forests. They cause floods and fires, soak up scant water supplies or choke irrigation canals. Some of the worst economic harm is done by introduced fish and molluscs. These can trigger the total collapse of marine ecosystems, taking away the livelihoods of thousands of people in some of the world's poorest countries.

So what can we do?

Tighten biological security, for starters. Only half the countries of the world have laws that begin to control the spread of invasive species. We also need to act fast if a plant or animal does sneak in under the wire. There's usually a brief honeymoon period when an invasive species can be identified and stopped in its tracks before it becomes a problem.

The Grand Cayman blue iguana has become critically endangered since the arrival of the invasive green iguana, from mainland Central America.

1 INVADERS IN COLD BLOOD

Today the pond, tomorrow the world. Reptiles and amphibians are among the oldest creatures on Earth, and these scaly and slippery invaders are surprisingly good at settling into new homes.

▲ *The spectacled caiman is a voracious predator that causes severe damage as an invasive species.*

Some of the most successful alien invaders on Earth are reptiles and amphibians. Yet many of these charismatic creatures are themselves at risk from the movement of plants and animals around the world.

Cold blood, warm heart

To qualify as a reptile, an animal must be cold-blooded, or ectothermic. This means that its body responds to the outside temperature, rather than being regulated from within. Reptiles bask in the sun to keep warm, and move into the shade or cool water when they overheat. A reptile should also have a thick, scaly skin so it retains water, and will probably have hatched from an egg. Animals ticking all these boxes are snakes, alligators, crocodiles, turtles, tortoises and lizards.

Amphibians don't have scales but have moist, often slippery skin. Think frogs, toads, newts and salamanders. Most grow from jelly-covered eggs into tadpoles and only move on land as adults – although they still spend much of their time in the pond.

Pets running wild

Reptiles and amphibians have been crawling and hopping across the planet for about 300 million years, long before those upstart dinosaurs made their brief appearance. Some of them have been remarkably adaptable, surviving ice ages and meteor collisions. This evolutionary ability to roll with the punches is why today you find huge Nile monitor lizards sunbathing behind Florida condos, cane toads from Latin America wreaking havoc in Aussie gardens, and tropical turtles relaxing on canal banks in the English Midlands.

The blame for their global spread lies mostly with the pet trade. Last count, 4.8 million households in the United States alone had reptilian pets, their popularity hugely boosted when the Teenage Mutant Ninja Turtles leapt on to our TV screens in the 1980s. Reptiles and amphibians also make appealing pets for people who live in crowded cities, as they don't need a big garden or taking for a walk in the park. The babies are able to live independently as soon as they hatch, so are sold while they're cute, wriggly and manageable. However, as they grow, they become less appealing flatmates, demanding more attention, more space and very often hogging the bathroom. Many are also potentially dangerous to small children.

One less than cuddly pet that may not have made it into our 100 worst offenders but is snapping at their heels is the spectacled caiman (*Caiman crocodilus*). This native of Central and South America is making a nuisance of itself everywhere from Florida to Cuba and the suburbs of Puerto Rico. It's the most abundant croc on Earth, highly adaptable compared with its relatives and able to steal food and territory from under their noses. It is also spreading a disease called caiman tongueworm to fish in Puerto Rico, and has a nasty bite. Not something you want to keep in the bathtub.

Frogs: the hopping barometers

Frogs are disappearing from ponds across the world. Nearly a third are threatened with extinction and as many as 200 frog and other amphibian species have vanished in the last three decades. That's a phenomenal rate – normally you'd expect only about four species to become extinct in 1,000 years. Their disappearance has a knock-on effect up and down the food chain. Frogs eat insects, including that old enemy of mankind, the mosquito. Frogs in turn are important food for fish and reptiles.

Reptiles and amphibians at risk

While some reptiles are successful invaders, many of the more specialised and less adaptable species are threatened by other exotic species. One such likeable but slow-moving reptile is the desert tortoise (*Gopherus agassizii*) of the US southwest, whose habitat has been engulfed by exotic grasses. These don't have as many nutrients as the native plants, so the tortoises go hungry.

Another reptile under threat is the handsome Tenerife speckled lizard (*Gallotia intermedia*). Critically endangered, it lives only on the northwest side of the island. Cats and rats have brought its population down to just 500 or so. The speckled lizard was only discovered in the mid-1990s – sadly, just a couple of decades later it may be about to disappear.

The tiny Mallorcan midwife toad (*Alytes muletensis*) is remarkable for its breeding behaviour: the males nurture the developing eggs on their backs. There are fossils of this toad dating back five million years, and it was thought to have become extinct during the Roman Empire. A small population was discovered on Mallorca in the 1970s, but today the toad is down to about 500 breeding pairs. It's under attack from the killer amphibian disease chytridiomycosis, massive habitat loss and predation from the introduced viperine snake (*Natrix maura*). However, this venerable amphibian now has masses of local and international support so still has a fighting chance.

◀ *Male Mallorcan midwife toad carrying eggs*

▶ *The Madagascar day gecko is now an invasive species in Florida.*

The extinction of these animals is a dire warning for life on Earth because they are nature's eco-barometers. They are extremely sensitive to pollution and other environmental hazards, so their disappearance indicates there's a serious problem. If frogs are hopping off the lily pad, in other words, the rest of us are likely to follow.

1. American bullfrog

Lithobates catesbeian

When chefs put American bullfrogs on menus around the world, they created huge survival problems for other species. The American bullfrog is a bruiser of an amphibian. It weighs in at two-thirds of a kilo, measures nearly half a metre nose to toetips at full stretch, and will eat anything that moves and is smaller than itself. That means it dines on other frogs (including smaller bullfrogs), birds, snakes, mammals and fish, as well as insects and other creepy-crawlies. In its native range it has plenty of natural enemies, but few predators have developed a taste for it in its adopted territory.

Big bully

Numerous bullfrogs were exported around the world for the frog-leg industry, but soon hopped out to colonise local ponds and waterways. This species lives in swamps and ponds but adapts to a wide variety of watery homes – particularly manmade environments such as millponds or reservoirs where humans have already disturbed the natural ecology.

Not only is the American bullfrog a hungry predator, but it is also a big factor in the spread of the deadly fungal disease chytridiomycosis. This infectious epidemic is responsible for devastating declines in many species of amphibians around the world. The bullfrog can carry the fungus without developing the disease.

The bullfrog is native to the US southeast and some of its deadliest advances have been in the western United States. Here it has rampaged through populations of the Pacific chorus frog (*Pseudacris regilla*), the Californian tiger salamander (*Ambystoma californiense*) and the Oregon spotted frog (*Rana pretiosa*). Across the Big Pond, the American bullfrog has been blamed for the decline of amphibians in Germany, France and Italy.

Frog features

This frog is also a threat in Britain, with several sightings and evidence of breeding in southern England. Occasionally, bullfrog tadpoles are accidentally imported in consignments of water plants, but while the British haven't developed a taste for frogs' legs, some bullfrogs were deliberately imported as potential biological controllers of bugs and as living garden ornaments. So don't be tempted to add a few supersize froggy decorations to your pond – stick to gnomes.

Fact file

- **The adult bullfrog can leap up to an amazing 2m.**
- **A group of frogs is called an army.**
- **A female can lay up to 20,000 eggs at a time.**
- **This mammoth frog now occurs in an astonishing 40 countries across four continents.**
- **You can tell males from females by their larger eardrums.**

2. African clawed frog

Xenopus laevis

Many post-war pregnancies were discovered by means of the African clawed frog. It's one of the most common lab animals in the world and provided the basis of a clever pregnancy test in the 1940s, after scientists discovered the frogs would start laying eggs when injected with the urine of pregnant women. The African clawed frog was also popular in medical research because it is highly resistant to disease and has an antiobiotic called magainin in its skin.

Rough, tough and mean

The environmental problems really began when easier pregnancy tests were developed, and labs released the now-redundant African clawed frogs into the wild. The same survival traits that made this animal such a successful candidate for experiments allowed it to flourish in new habitats around the world. It can survive in water of up to 40% salinity and widely varying temperatures. In a dry season, it copes by digging deep into the mud and breathing air through a narrow tunnel, and can also migrate long distances between ponds. It's also a prolific breeder: a female can produce as many as 27,000 eggs.

The world is not enough

It would seem that planet Earth may be just the beginning for this fearsome frog. On 12 September, 1992, several female African clawed frogs were rocketed into outer space on Nasa's space shuttle *Endeavour*. The experiment was part of research into the effects of gravity on fertility and embryos. It was an uncomfortable twist on the classic sci-fi story of an alien invader hurtling through space – this time we might be the ones sending our own dangerous alien to colonise new worlds.

Not only is this species very adaptable, but it is also highly aggressive and will devour anything, dead or alive, including invertebrates, small fish and other amphibians – along with their eggs and larvae. In new territories such as California, it is likely to threaten the survival of endangered native fish such as the unarmored threespine stickleback (*Gasterosteus aculeatus*) and the tidewater goby (*Eucyclogobius newberryi*). It has also been blamed for a decline in numbers of the California red-legged frog (*Rana aurora draytonii*).

As if that weren't bad enough, this frog is another carrier of the dreaded chytridiomycosis disease. The very first case was discovered in an African clawed frog in the late 1930s, when the species was already being exported worldwide.

Fact file

- The African clawed frog doesn't have a tongue.
- In African folk medicine it is used as an aphrodisiac.
- It's the only frog with clawed toes.
- In the 1960s, it became the first vertebrate animal to be cloned.

3. Cane toad

Bufo marinus

The cane toad is the poster boy of invasive species – squat, ugly and a master of chemical warfare. Introduced to many countries, this South American land toad is globally notorious following its colonisation of Australia. For such an apparently slow-moving animal, it's expanding its territory at an incredibly fast 50km annually. Worryingly, new research indicates the spread is speeding up every year, and some tracked toads crawl a kilometre each night.

Deadly dinner

This is a hefty 2.5kg amphibian with dry, warty skin. Ugly or not, however, the cane toad is a remarkably good breeder. The females can lay a whopping 30,000 eggs at a time and adults survive for up to 15 years. Cane toads are also remarkably adaptable. They can tolerate a wide range of temperatures, from a chilly 5°C to a steamy 40°C. Dehydration is no problem either: they can withstand the loss of up to 50% of the water in their bodies and are making themselves at home in the Queensland desert – not bad for an animal that came from the South American rainforest. Today, there are about 200 million cane toads in Australia.

And that means trouble. Because not only are cane toads blamed for spreading diseases to humans, such as salmonella, but they are also poisonous to pets and wildlife. Even the tadpoles are toxic.

Australia's most wanted

This unwelcome amphibian is the most famous Australian outlaw since Ned Kelly, introduced to the only inhabited continent on Earth that doesn't have its own native toad. Millions of dollars have been spent trying to eradicate the cane toad, with little success. The Western Australian government even called in the army in a bid to halt the toad advance. Rural communities have been given grants to build toad fences around wildlife reserves. In 2007, a Queensland landlord offered two beers for every cane toad brought into his pub. The idea caught on – two years later, the state staged a mass cull called Toad Day Out. Volunteers caught and bagged toads to be weighed then humanely killed. The event was the brainchild of a Queensland politician, who has been campaigning for the Australian government to pay a bounty for captured cane toads.

The biggest losers have been native predators, poisoned by the cane toad's spray, a nasty chemical called bufotoxin. Species like goanna lizards (genus *Varanus*), freshwater crocodiles (*Crocodylus johnsoni*) and the cute, cat-like quolls (genus *Dasyurus*) have no inbuilt defences against the toxin. Goanna numbers are reported to have dropped by 90% in areas that are teeming with cane toads.

Some birds, such as the Australian ibis (*Threskiornis molucca*) and black kite (*Milvus migrans*) seem to be immune to cane toad venom. Scientists believe they may have built up natural

▲ *Cane toads may capture and consume small mammals.*

resistance through long dealings with poisonous amphibians in Asia. Helping the cause somewhat is the fact that cane toad tadpoles regularly snack on cane toad eggs.

When the solution becomes the problem

It's called biological pest control – letting one organism loose to get rid of another. Although there have been some successes, many efforts have caused ecological chaos. Cane toads were shipped around the tropics around the early 20th century, and in the 1930s about 100 were brought from Hawaii into Queensland by the Bureau of Sugar Experiment Stations. The idea was to use the toads as biological bagmen to get rid of French's and greyback cane beetles, which were destroying sugar-cane crops. Unfortunately, the toads preferred different prey, from other amphibians to all sorts of water and land insects, snails and even small mammals.

Now scientists are looking at biological controls such as viruses to get rid of the cane toads, though some environmentalists are concerned about unleashing yet another alien invader into the Australian bush. Another possibility is pheromone gene therapy, as a form of population control.

Fact file

- **The toad's toxin was used as an arrow poison by native South American forest people.**
- **The toad is known as 'spring chicken' in Belize, and is eaten in Latin America by people who know how to remove the poisonous parts.**
- **Bufotenin, one of the chemicals in the skin of the cane toad, is classified as an illegal drug under Australian law; the Japanese use it as hair restorer.**
- **The cane toad loves eating honey bees and is a serious pest for Hawaii's and Australia's beekeeping industries.**
- **In Hawaii, about 50 dogs die each year after eating cane toads.**

4. Caribbean coqui frog

Lithobates catesbeian

The Caribbean coqui frog is giving the people of Hawaii sleepless nights, because of the male's loud and incessant mating call. It's not an unpleasant sound, resembling the fluty notes of some tropical bird, but delivered at high volume and all night long without pause, it can quickly become an irritant. From an ecological viewpoint, though, singing too loudly is the least of the coqui's crimes.

Loud and hungry

The coqui hopped across the Pacific from its native Puerto Rico back in the 1980s, probably stowing away on ships introducing exotic nursery plants. With no natural diseases to keep its population in check – and perhaps because of the Latin American serenading – numbers soon grew to a level where there are now as many as 20,000 frogs per hectare in Hawaii. And all partying hard through the night.

This dainty little frog has hopped into a whole lot of trouble with environmentalists since its arrival, eating numerous endangered insects, spiders and snails. A single hectare's worth of frogs will devour 114,000 invertebrates each night. That reduces the food supply for Hawaii's native bug-eating birds, many of which are in big trouble anyway because of habitat loss, and threatens endemic plants that rely on insects for pollination. In an ironic twist, the coqui frog has also become a readily available snack for other seriously invasive aliens, such as rats, snakes and the Indian mongoose (*Herpestes javanicus*).

When people can't sleep they get really, really annoyed. The Hawaiians are fighting back with anti-coqui chemical trials. They've had modest success on outer islands by spraying breeding areas with citric acid, which clears the frogs without harming native species. Meanwhile, so irritated are the sleepless Hawaiians with coquis that anyone caught transporting or releasing these frogs faces a fine of up to US$200,000 – and three years in prison.

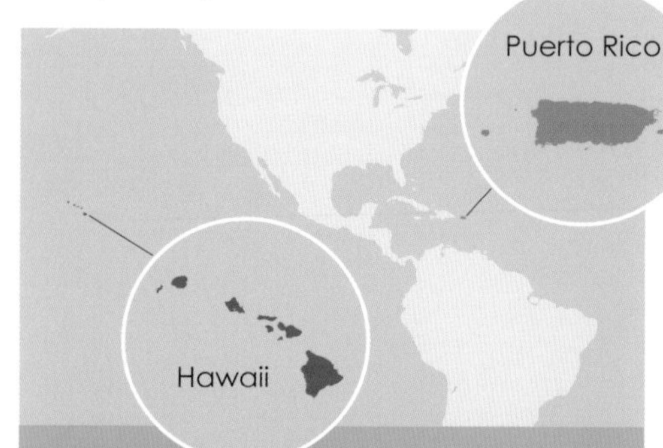

Fact file

- **The coqui frog breeds on dry land and doesn't go through a tadpole phase, but emerges from its egg as a fully formed (though tiny) froglet.**
- **It has pads on the ends of its feet so it can stick to surfaces, but no webbing between the toes, making it an incompetent swimmer.**
- **The male's mating call reaches up to 90 decibels, about the same noise level as a lawnmower.**
- **As if the coqui wasn't bad enough, Hawaii is suffering from another invasive Caribbean alien, the greenhouse frog (*Eleutherodactylus planirostris*).**

5. Nile monitor

Varanus niloticus

This huge lizard's bid for world domination all started with a single pregnant female, who crawled out from captivity and kick-started a new population. Now, ever since the early 1990s, these two-metre-long lizards have been regularly spotted sunbathing in backyards around the Cape Coral region in Florida.

Monitoring progress

The Nile monitor is native to Africa, as the name suggests. But today it is spreading its territory to the islands of Pine Island, Punta Rassa and Sanibel around the Florida coast. So concerned are the local wildlife authorities about the destructive impact of these big reptiles that they have passed a law charging owners an annual US$100 to keep one. Only adults can own the monitors and they must be microchipped (the lizards, not their owners).

The key to the success of Nile monitors, as with many invasive reptiles, is their adaptability. These animals enjoy an amphibious lifestyle. In the water they can swim like fish – using their long tail as a paddle – and remain hiding below the surface for up to an hour. On land they can climb trees, scale walls like Spiderman, sprint at up to 25km/h and tunnel far below ground to escape from predators. Monitors are also excellent breeders, laying about 60 eggs every two years, with a good survival rate.

The big problem comes with this reptile's diet. Nile monitors will devour almost anything that crosses their path, including roadkill and turtles complete with shells. In the Cape Coral region they have become a serious threat to the burrowing owl – a Florida Species of Special Concern. With only about 1,000 pairs of owls in the area (the densest population in the world), the stakes are high. Ecologists are also concerned about the Nile monitors raiding the nests of sea turtles, sea birds, American crocodiles and gopher tortoises, and preying on the rare Sanibel rice rat (*Oryzomys palustris sanibeli*).

Fact file

- Nearly half of the approximately 60 monitor species in the world are kept as pets in the US.
- Monitors are difficult to tame so are often set free by owners.
- The Nile monitor is the longest lizard in Africa, reaching a maximum of 2.6m.

Those attempting to rid Florida of this menace have their work cut out. 'Do not attempt to apprehend,' would probably be the police advice. Nile monitors have lethally sharp claws, a whip-like tail that they lash at assailants and powerful jaws dripping with pathogen-laden saliva. It is worth remembering that their big cousin, the Komodo dragon of Indonesia, is a known man-eater.

6. Green iguana

Iguana iguana

In the early 1990s, two large green or common iguanas were kept as class pets by pupils at the George Town Primary School on Grand Cayman Island. Within just a few years, isolated green iguanas started to appear in backyards and roadsides. The school pets had laid eggs and the babies had managed to sneak out of the cage and colonise the island.

Time out

If the island authorities had been alerted to this breakout in the early stages, they might have been able to round up the South American green iguanas, ban their future import and stop their spread. Unfortunately, scientists found out too late to stop a mass invasion.

Within a remarkably short time, the incomers had colonised mangrove canals and swimming pools – much to the disgust of the islanders. They thrived, multiplied and soon vastly outnumbered the endemic and now critically endangered Grand Cayman blue iguanas (*Cyclura lewisi*). By 2006, the invasive aliens were breeding right across the island.

Surviving on a tropical island was a walk in the park for the alien iguanas, used to a much tougher environment in the rainforests of Central and South America. They are agile, adaptable and very capable of defending themselves. Even so,

The blue brothers

The Grand Cayman blue iguana is one of the longest-living species of lizard, with a lifespan of 60 years. It's also one of the most critically endangered animals on Earth. Scientists studying lizard fossils believe it was common across Grand Cayman until Europeans settled two centuries ago.

By the late 1980s the wild population of blue iguanas was down to just 15 animals, thanks to predation from cats and dogs, and large-scale habitat loss. With the arrival of competing green iguanas, things looked very bleak for the blues – and for the balance of nature on Grand Cayman, as blue iguanas are the most effective means of dispersing the seeds of endemic plants. However, there is hope. In the early 1990s, a captive breeding programme was set up by environmental scientist Frederick Burton through the National Trust for the Cayman Islands. Today, the Blue Iguana Recovery Programme has seen more than 200 captive-bred blue iguanas released into protected reserves.

in their native territory, their numbers are kept in check by large birds of prey and jungle cats. Sadly, the only jaguars in Grand Cayman's banking district have four wheels.

Florida invasion

Green iguanas stowing away on banana boats are sneaking into North America and are now a nuisance throughout southern Florida as far south as Key West. Their numbers have been boosted by released pets, as owners watch the cute little lizards they bought from the shop grow into 2m monsters weighing in at about 5kg and with razor-sharp teeth.

Adult iguanas can be aggressive during their breeding season and require a huge amount of space – virtually their own room. Used to humans, illegally released semi-wild individuals tend to hang around the Florida suburbs, ripping through flower beds, fouling pools and invading cellars and attics when the weather cools.

The potential impact of these lizards on Florida's natural environment is even more serious. Green iguanas take over the burrows of Florida's rare burrowing owl (*Athene cunicularia floridana*). They also eat native plants that are the main food for the Miami blue butterfly (*Cyclargus thomasi bethunebakeri*), one of the most endangered insects in the United States.

World domination?

The march of the green iguana continues. It is a nuisance animal on Puerto Rico and has invaded the Rio Grande Valley in Texas. Now it's wild in Hawaii's Oahu and Maui islands. Such is the concern about green iguanas' impact on Hawaii's fragile ecosystem that it's not only illegal to keep iguanas as pets but anyone importing them may face a hefty fine and even jail.

Fact file

- **Green iguanas are eaten by people in Central and South America who claim they taste like chicken.**
- **They spread the disease salmonella to humans.**
- **Although they are notoriously difficult to keep, they remain the most popular reptile pet in the United States.**
- **They are vegetarian.**
- **Green iguanas lay as many as 70 eggs in a clutch.**
- **A new breeding colony on Anguilla may have floated on driftwood rafts some 320km from Guadeloupe.**

7. Veiled chameleon

Chamaeleo calyptratus

The ability of chameleons to blend into their background makes them supremely adaptable animals and thus highly successful alien invaders. This particular species hails from Saudi Arabia and Yemen, hence its alternative name of Yemen chameleon. Thanks to the pet trade, however, it has now taken a scaly foothold in Hawaii and Florida.

Versatile reptile

Chameleons are masters of disguise. Green, blue, orange, red or yellow, black or white: you name it, there's a chameleon shade to match. This makes them very hard to catch – as do their conical, swivelling eyes, which work independently of one another to spot predators both in front and behind. They are also expert tree climbers, with a prehensile tail and tong-like toes that allow them to manoeuvre securely among the flimsiest of branches.

These versatile lizards have a highly adaptable diet. They are voracious omnivores that will happily devour everything from small birds, mammals and insects to flowers, buds and leaves. This catholic appetite can prove disastrous in fragile habitats such as the Hawaiian Islands. Not only do the chameleons consume many native species, but they also compete against the local fauna for any available food.

Veiled chameleons are equally versatile in their habitat preferences. In Hawaii, they flourish along the dry coastline, as well as in deep forest and on mountains up to 3,500m. Their ability to breed fast, laying up to 90 eggs in a clutch three times a year, helps them colonise new territories quickly, and they can live for up to eight years in the wild.

All this helps explain why Florida's population of veiled chameleons, descended from pets released into the wild, is now moving steadily across the state, invading the Everglades and the area around Fort Myers. The Florida Conservation Authority is trying to halt further introductions by running amnesty days, on which chameleon owners may surrender their exotic pets rather than release them into the wild. The Hawaiians have gone a step further, having made it illegal to keep veiled chameleons as pets at all. Anyone trading in these destructive aliens faces a fine of US$200,000 and a possible prison sentence, so concerned are the people of Hawaii about the threat to their precious native wildlife.

Hawaii

FLORIDA

Fort Myers

Fact file

- The veiled chameleon is the most commonly bred of the 80 species of chameleon in the pet trade.
- Like other chameleons, it has a sticky tongue that can shoot out one-and-a-half times its own body length to capture prey.
- Chameleons can sleep suspended upside down from a branch.

8. Brown tree snake

Boiga irregularis

When the lights go out on Guam, an island territory of the US in the western Pacific, the locals blame the brown tree snake. This alien invader is an excellent climber that slithers along power lines and short-circuits electrical supplies, causing up to US$4 million worth of damage every year. Yet while it is a nuisance for the island community, the impact of this snake on Guam's wildlife is far more sinister. Since it was accidentally introduced after World War II, the brown tree snake has directly caused the extinction of more than half of Guam's native bird and lizard species, as well as two species of native bats.

Mounting death toll

The snake's arrival resulted in a huge environmental disaster. Ten forest bird species were silenced forever, and the plummeting numbers of pollinating birds soon caused a decline in native flora, damaging the island's whole ecology.

This Indonesian, Australian and Papua New Guinean snake has no natural predators on Guam. As a result, it grows much longer than usual in its adopted island home, reaching nearly 3m. Scientists also suspect that on Guam it breeds year-round, rather than seasonally as it does in its original forest territory. At its peak, there were nearly 5,000 snakes per square kilometre on Guam.

The brown tree snake is also a threat to human health, its venomous bite accounting for one in every 1,000 hospital admissions on the island. While it is only mildly dangerous to adults, its venom can seriously harm babies and small children. With superb climbing skills and well-camouflaged skin, the snake can sneak pretty much anywhere it likes, and can pass through very small openings into buildings, the holds of cargo ships and even into the wheel wells of military aircraft.

Now the United States has declared war. Since the early 1990s, the American Defense Department has spent more than US$5.4 million on snake control measures in Guam. They're also fighting to stop the spread of these snakes by checking all aircraft before take-off, to ensure these adaptable aliens don't hitchhike on military flights to Hawaii.

Fact file

- **Sniffer dogs are trained to detect brown tree snakes around ports and airports.**
- **Young brown tree snakes up to 1m long feed exclusively on lizards.**
- **Predation by brown tree snakes has devastated Guam's poultry industry.**

9. Burmese python

Python molurus bivittatus

Lock up your kids and hide away your pets. Check the pig pen and the chicken coop – a large python is on the loose and it's moving across Puerto Rico and southern Florida. This nocturnal predator has been known to overcome full-grown alligators and is a serious threat to many smaller native species.

Squeezing out the natives

The Burmese python likes to eat small mammals and birds but will swallow anything it comes across, from frogs, toads and lizards to bats and even other snakes. In Puerto Rico it out-competes the island's two native constricting snakes, the endangered Puerto Rican boa (*Epicrates inornatus*) and the Mona Island boa to (*Epicrates monensis*). And in the Florida Everglades these bold snakes will prey upon other formidable hunters, even as large as bobcats and racoons.

The Burmese python is one of the biggest snakes in the world. Stretched out it can reach more than 6m, with a body thicker than a grown man. It is an Olympic-class swimmer and can hold its breath underwater for half an hour, but also slithers up trees and across the ground at speed. This water-loving serpent prefers marshes around tropical forests, but constant pressure from humans means it's just as likely to turn up in your swimming pool. Wherever it is, this alien invader needs a lot of time in the water, particularly just before it begins to shed its skin.

There are also plans to block canals and levees, to prevent the further spread of these dangerous aliens into the national parks and northwards to other parts of North America. It's a race against time. Burmese pythons are fast breeders, with the females laying as many as 100 eggs in a clutch. Owning a python is still not illegal, although anyone keeping one of these snakes in Florida must now pay US$100 a year for a licence.

And the reason for its introduction to Florida and Puerto Rico? The pet trade – of course! More than 144,000 Burmese pythons were imported into the US in just the first five years of the 21st century. Cute little snakes are less appealing as they grow into huge potential man-eaters, and adult snakes are regularly set free into the wild. Today there are an estimated 30,000 Burmese pythons in the Florida Everglades National Park.

Slithering further afield

Until just a few years ago, scientists believed tropical snakes were unlikely to slither further north and become a problem. However, the Burmese python is proving to be remarkably adaptable, able to find shelter during the cooler months and food throughout the year. Today, biologists believe pythons will be able to colonise a third of the United States, as far west as California, by the end of this century.

Fighting back

Enter the Python Patrol. This is a squad of volunteers set up by the Florida Nature Conservancy to help track down these alien snakes. These volunteers are people who spend time outdoors, such as postmen and road crews, and are likely to spot Burmese pythons sunbathing on the warm tarmac. A phone hotline has been set up to report findings, then wildlife teams take over using radio tracking devices and trained sniffer dogs.

Fact file

- **This snake has poor eyesight, and stalks prey using chemical receptors in its tongue and heat-sensors along the jaw.**
- **It lives for up to 25 years in the wild.**
- **The world's biggest Burmese python was the misnamed 'Baby' at the Serpent Safari Park in Illinois. Baby weighed 183kg and was 8.23m long.**
- **Does your pet python stretch out on the sofa next to you? He's not feeling cuddly – just measuring to see if you're small enough to eat.**
- **A Burmese python is able to overpower and kill an adult human once it reaches 5m long.**

Protecting the Galápagos

The Galápagos Islands lie 1,000km off the coast of Ecuador, and are famed as the setting where Charles Darwin developed his theory of evolution. In the five weeks Darwin spent exploring these islands in 1835, he became fascinated by the diversity of the wildlife and by the way animals that appeared to be of the same species looked and behaved differently from one island to the next. Sadly, the bird which first inspired Darwin's ideas became extinct just 50 years after the voyage of the *Beagle*.

▲ Floreana mockingbird on Floreana Island.

Making a mockery

In particular, Darwin noticed differences between the Floreana mockingbird (*Mimus trifasciatus*) on Floreana Island (known in those colonial days as Charles Island) and mockingbirds he'd collected on San Cristobal, 80km across the sea. Darwin was astonished by how much these 'mocking-thrushes' varied between islands. It was a thought process that led to his startling conclusions about natural selection.

▼ Volcan Alcedo giant tortoise on Isabella Island.

The mockingbird on Floreana was killed off by a lethal combination of rampaging black rats (*Rattus rattus*) and the disappearance of its habitat as goats bulldozed their way across the island, clearing the birds' favourite food, the prickly pear cactus (*Opuntia megasperma*).

However, the mockingbird story may have an unexpectedly happy ending. Using detective techniques worthy of Hercule Poirot, scientists have uncovered genetic clues from Darwin's original collection that hold the key to reviving the Floreana mockingbird. A few hundred mockingbirds are nesting on the nearby Champion and Gardner islets, and the genetic studies revealed that they belong to the same species as those collected on Floreana by the *Beagle* crew.

Naturalists working for the Charles Darwin Foundation and the Galápagos National Park Service are now hoping to save this critically

The finch intervention

Rats, cats and a killer parasite mean a little brown bird called the mangrove finch (*Camarhynchus heliobates*) is hanging on to survival by a single claw. There are only about 100 or so of these endemic Galápagos birds left, stuck in three small patches of mangrove on Isabela Island. Enter the cavalry, in the form of ornithologists from the Durrell Wildlife Conservation Trust, the National Park Service and Charles Darwin Foundation. They began poisoning rats around the mangroves. Before the baiting began, fewer than one in five eggs survived – now more than one in three eggs is hatching.

endangered bird by removing rats, goats and other invasive species from Floreana, to make it safe once again for the birds.

Saving the giant tortoise

The ancient ancestor of the Galápagos giant tortoise (*Geochelone nigra*) would probably have fitted inside a shoebox. But without predators and competition for food in the islands it grew larger ... and larger ... eventually becoming the 250kg heavyweight we know today. Unfortunately, size was no protection against centuries of hunting by pirates and whalers, who killed as many as 100,000 tortoises. Then came the rats and cats, devouring the eggs and hatchlings. Three unique subspecies of Galápagos tortoise became extinct and a fourth will follow soon when Lonesome George dies. He's the last remaining Pinta tortoise, living out his retirement at the Charles Darwin Research Centre on Santa Cruz Island.

Something had to change if the tortoises were to be saved. The National Park Service began a clearance programme called Project Isabela to rid North Isabela Island of the thousands of goats that had destroyed huge swathes of forest. The results were startling: as goat numbers dropped, the forest began to regenerate naturally. Not only were the tortoises back in business, but the Galápagos rail (*Laterallus spilonotus*) that hadn't been seen on the island for decades began returning to nest in large numbers.

Today there are about 15,000 tortoises in the islands, and their prospects are improved. Scientists are tackling invasive aliens, and a captive breeding programme for tortoises is proving successful.

Fact file

- **The Galápagos Islands are globally recognised as a key area for biodiversity and 95% of that original diversity still exists.**
- **Half of the islands' endemic species are under direct threat because of competition from invasive species.**
- **Invasive species brought in by people form the main threat to the biodiversity and economy of the Galápagos Islands.**
- **New research shows that the mangrove finch may be 'splitting' into two new species, which means that both will be in even more danger because of their tiny populations.**
- **Visitors to the Galápagos may unwittingly bring alien plant seeds to the islands on their clothes and bags.**

Even saving the mockingbirds and finches will help the giant tortoise, as these little birds eat the ticks that drive the reptiles potty.

Safeguarding the plants

There are more than 700 species of introduced plant in the Galápagos today compared with fewer than 500 natives, some 180 of which are now under threat. Scientists are studying ways to control invasive weeds such as blackberry (see page 147).

▼ *Feral goats on Santiago Island have turned dense forest into grasslands.*

10. Red-eared slider

Trachemys scripta elegans

Red-eared sliders – or terrapins – are sometimes given away as fairground prizes. Unfortunately, they cause environmental problems in their adopted territories long after the fair has left town. These freshwater turtles native to the southern US states are named for the distinctive red stripe along the side of their head. The babies look a bit like clockwork toys, and appeal to children as pets.

Turtle power

Concerns were raised about these turtles as far back as the 1960s when the American pet trade began exporting them to Europe, New Zealand and Australia. They became a global problem in the 1980s when the hit cartoon *Teenage Mutant Ninja Turtles* saw red-ears morph from hoopla prizes to must-have pets everywhere from Melbourne to Moscow. Between 1989 and 1997,

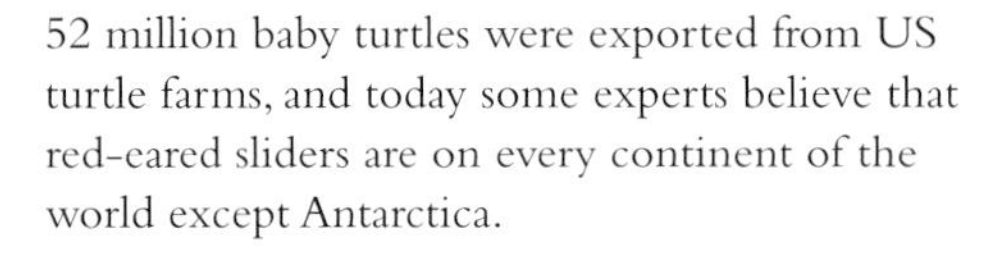

52 million baby turtles were exported from US turtle farms, and today some experts believe that red-eared sliders are on every continent of the world except Antarctica.

Unfortunately, babies the size of 50 pence pieces soon grow to be large and rather smelly dinner plate-sized adults. Mothers the world over weren't amused, and that's why there are now so many sliders in canals, rivers and city ponds.

These highly efficient omnivores devour plants, small water creatures and even small snakes that would normally be the food supply for native turtles. In Europe, they not only compete for food against the less aggressive European pond turtle (*Emys orbicularis*), but shove their European cousins away from the best sunbathing spots. In the US, they threaten rare species like Pennsylvania's red-bellied turtle (*Pseudemys rubriventris*), and the endangered and endemic Pacific pond turtle (*Clemmys marmorata*) in Washington State. Baby red-ears hatched in turtle farms are raised in cramped conditions, and may pass diseases and bacteria to wild species.

Red-ears can also pass salmonella to children. This health problem saw US health authorities ban the sale of turtles smaller than 10cm back in the 1970s. However, you can still buy tiny red-eared sliders on the Internet in the United States. The EU bans importation, although existing owners may keep their turtles. The Australians are a lot tougher. Anyone found harbouring a red-eared slider, Ninja or otherwise, faces up to five years behind bars.

Fact file

- The red-eared slider likes freshwater lakes with muddy bottoms, lots of plants and rocks or logs for sunbathing.
- It can live for up to 50 years in captivity.
- Slider eggs need to be incubated at 25°C for 60 days, which is rarely on offer in the UK. However, there are fears that climate change could provide the opportunity for a population explosion.

11. Common snapping turtle

Chelydra serpentina

It strikes with the speed and ferocity of a rattlesnake, crushing its prey with lethal mandibles and powerful jaws. Even in the snapping turtle's home territory there are few predators willing do to battle with this aggressive beast that can weigh as much as a five-year-old child (even though it's only the length of a new-born infant).

All you can eat

The snapping turtle is a native of the United States, east of the Rockies then from southern Canada down to Ecuador. It eats water birds, snakes, large fish, crayfish, insects and worms and is notorious for dining on its near-relatives – other, less ferocious, turtle species. Tough, long-lived and able to shuffle long distances when necessary, this is not an animal you want to invade the rest of the planet.

Very popular and much loved in its native areas, this beefy beast was probably moved to its new territories by the pet trade. Like other turtles it is very appealing when small, but grows into a formidable and sometimes ill-tempered adult. Letting a problem pet loose into the wild might seem like an easy solution, but its consequences are frequently disastrous for local wildlife.

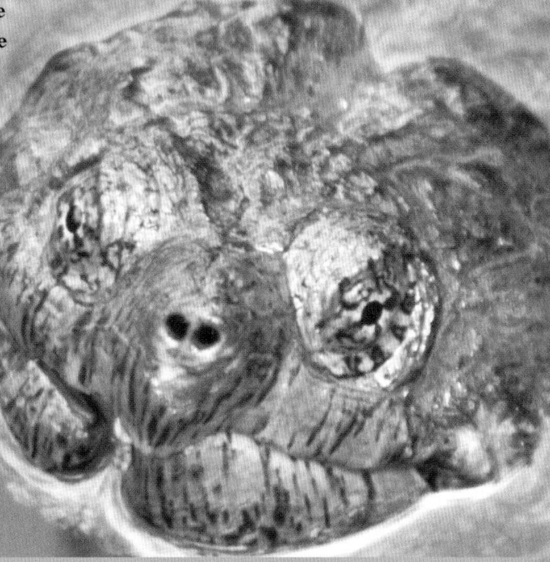

The first stop on the snapping turtle's world tour was California and Oregon, where it is now strictly outlawed. In Oregon, it is posing a serious threat to the survival of the critically endangered and endemic western pond turtle (*Actinemys marmorata*) and western painted turtle (*Chrysemys picta bellii*). Not only is the snapping turtle preying on its passive cousins, but it is also competing against them for food and the sunbathing or basking sites necessary for a turtle's well-being.

Fact file

- The snapping turtle can survive in a wide range of habitats, from lakes to farm ponds, wetlands and rivers.
- It may live for 40 years, and eventually grows so fat that it can't move easily on land.
- The power of the snap is enough to easily amputate a careless human finger.

Cold snap

The spread of snapping turtles doesn't end in the Americas. They've also been released into the wild in the UK and the Netherlands. So far sightings have been of isolated individuals, which are usually caught when they are spotted in city ponds or canals. However, scientists are concerned that snappers may pose an even bigger threat to native amphibians than the more widespread (at the moment) red-eared slider because they adapt better to cold climates, and could breed in countries as cold as Siberia.

2 FURRY FIENDS AND FELONS

A suite of unique characteristics makes mammals the most advanced members of the animal kingdom. Unfortunately, their ability to adapt to the most extreme conditions has also turned many species into deadly and destructive invaders.

The red fox is a devastating predator wherever it has been introduced.

The most successful invasive species on Earth by far is a mammal – good old *Homo sapiens*. Humans aside, our fellow mammals are generally more adaptable than other types of animals – several species are able to thrive almost anywhere on Earth. Traits such as the ability to control their own body temperature, and the fact that females carry around an in-built supply of nutritious food for their babies – in the form of milk – help equip them for making a living wherever they choose.

Expert invaders

The mammal species that have prospered most as invasive aliens – cats, rats, mice, pigs, goats – have several things in common. They are generalist eaters, with lots of strings to their bows when it comes to ways of finding lunch. They are also intelligent and opportunistic. And a high birth rate means they can build their numbers quickly, even if their life span is briefer than that enjoyed by other animals. Perhaps most importantly, they have long-standing associations with

▲ *A feral dog on Isabella Island, Galápagos*

▼ *Feral zebu cattle in Costa Rica cause serious damage by overgrazing.*

Man's best friends

We have domesticated many mammal species over the years. We eat them, drink their milk, cut off their hair to weave clothes, use them as vehicles, train them to work for us, and sometimes we just like cuddling them. The species which we were most successful at domesticating were those that were already prospering in the wild when we found them – mammals that were hardy, numerous and able to adapt easily to new ways of life. Being looked after by people might sound like an easy life, but many animals are simply too delicate, specialised or highly strung ever to be domesticated.

As we explored the world over the last three or four centuries and expanded our human territory, we brought our animal companions with us. They provided portable fresh meat, labour and rodent control, and no-one guessed (or perhaps no-one cared) that they could do extraordinary harm to native wildlife and ecosystems elsewhere in the world. Today, we have learned our lesson to some extent, and many countries have very stiff penalties when it comes to liberating non-native domestic mammals, but ignorance is still rife. A resurgence in the popularity of unusual pets, from Australian sugar gliders to African pygmy hedgehogs, in the western world could lead to more invasive mammals making themselves at home thousands of miles from their native lands.

humans, whether as our friends or our persistent exploiters, and thanks to these associations they have travelled the world in our company.

Island invasion

Invasive mammals on islands form one of the biggest threats to our planet's biodiversity. Islands are home to a disproportionate number of animal species that have evolved alongside only a few predators, if any. They simply don't have the defences to cope with an onslaught of hardened mammalian invaders from elsewhere. Of the 724 known animal extinctions during the last four centuries, at least half were island species. Predators such as rats have decimated seabird breeding colonies on islands across the globe. Three-quarters of seabirds are now threatened by invasive species. Scientists believe removing invasive mammals from islands is one of the most important steps we can take to prevent future extinctions of any other species.

Continental crisis

While island species are more vulnerable to invasion by mammals, there is still much to be done limiting invasive species on continents as well. American minks (see page 38) are decimating the population of native European minks, while coypus (page 42) are upsetting the balance of nature in European rivers. Some of the most catastrophic losses have been in Australia where native species – including mammals – are unable to cope with the arrival of more voracious predators from harsher climates, such as the red fox (page 40). One such Aussie mammal that has been chased to the brink of existence is the numbat (*Myrmecobius fasciatus*).

Feral mammals also cause massive economic losses. In the US, it's been estimated that feral pigs cause at least US$800 million worth of damage every year. Not all of the nuisance mammals on continents are ravaging carnivores.

▲ The Chinese water deer is now established in the UK.

Fact file

- There are about 4,260 species of mammal.
- New mammals are still being discovered, such as the Borneo clouded leopard (*Neofelis diardi*) and the kipunji (*Rungwecebus kipunji*), a monkey from the highlands of Tanzania.
- Rodents comprise the most diverse mammal group: four in every ten mammal species belong to the order Rodentia. In all, there are 2,277 species, including rats, squirrels, chipmunks, lemmings and beavers.
- Bats are the second-largest order of mammals, clocking in with 900 different species.
- Mammals appeared on Earth about the same time as the dinosaurs.

▲ Scientists have not yet resolved whether Borneo's elephants are indigenous or descended from captive elephants presented to the Sultan of Sulu in 1750.

12. Indian musk shrew

Suncus murinus

The Indian musk shrew may weigh less than a slice of bread, but it advances across new territory with the speed and force of a Sherman tank. Five years after a few shrews were seen scuttling about Apra harbour on Guam in the early 1950s, the Indian musk shrew had become common across the entire island, devouring insects and plants wherever it went. It has now been blamed for the decline and possible extinction of the island's population of pelagic geckos (*Nactus pelagicus*).

Taming of the shrew

This opportunisitic shrew began life in the forests of India, but soon discovered easier pickings around human habitations. It likes to live around houses, scuttling along the walls at night or snoozing in piles of swept-up leaves in the courtyards. By hitching rides on ships, carts and, later, motorised vehicles, it extended its range far beyond India to the Middle East in the west, Japan in the east and beyond to the Indian Ocean and Pacific islands. It is more adaptable than European shrews and breeds rapidly, the females producing two litters a year.

The rapid spread and impact of the Indian musk shrew shows that even the smallest invaders can become a biodiversity threat of global significance. The dodo won't be the last species to slip into extinction on the island nation of Mauritius, if scientists don't find ways to control this tiny invader.

Mainland Mauritius has been home to the shrew since the mid 18th century. However, just a decade after its arrival on the offshore island of Rodrigues, scientists noticed the disappearance of a native centipede. More invertebrate species have since vanished. The shrews not only compete with local species for food, but also eat small endemic animals and destroy plants by digging up the roots. Eradication programmes have been attempted, with only local success on Indian Ocean islands, the shrews being much less vulnerable to bait than rats and mice.

Comores
Sri Lanka
Mauritius
Reunion
Maldives

Fact file

- **The musk shrew appeared in Rudyard Kipling's *The Jungle Book*. He mistakenly called it a 'musk rat', and an illustration of that American animal appears in the book.**
- **It is known as a chuchunder in India, and is tolerated because it eats cockroaches and other unwanted house invaders.**
- **Glands on the side of the musk shrew's body produce a strong scent, which puts predators off.**

13. European hedgehog

Erinaceus europaeus

It is a welcome visitor in an English country garden, foraging in the rose beds to keep the slugs under control. Yet ship the hedgehog to an island and it will turn into a voracious predator, capable of seeing off indigenous species faster than you can say Mrs Tiggy-Winkle.

▲ A hedgehog tucks into a clutch of lapwing eggs.

Prickly problems

The hedgehog enjoys a varied diet. It eats slugs and other creepy-crawlies but also consumes lizards, frogs, eggs and baby birds with equal gusto. It breeds well, and by hibernating can adapt to relatively harsh climates.

Following the introduction of the hedgehog to the Outer Hebrides in Scotland back in the 1970s, to keep garden pests under control, its population grew to 5,000 individuals. Unfortunately, these hungry hogs developed a taste for eggs and became a serious threat to the survival of wading birds such as dunlins, ringed plovers, redshanks and lapwings, which breed along the sandy coastal grasslands. Their chicks and eggs made easy pickings for hedgehogs. On the islands of North and South Uist and neighbouring Benbecula, the number of shoreline birds had been almost halved by the turn of the century.

A scheme to cull the animals sparked outrage from the hedgehog-loving British public. As a result, hundreds of hedgehogs on Benbecula and North Uist were rounded up and shipped to shelters on the mainland. On South Uist, naturalists fenced off bird breeding sites to protect them from hedgehogs. The hedgehog has now all but disappeared from North Uist and Benbecula, but is still thriving on South Uist. As the islands are linked by causeways, naturalists are concerned that the invaders will waddle their prickly way across the sea to cause more destruction.

In New Zealand

Hedgehogs are rolling into a whole lot of trouble in other parts of the world. In New Zealand, they compete with native insect-eaters such as kiwis. Here they are being culled, and vulnerable areas of rainforest are being protected with hedgehog-proof fences. There are no native hedgehogs in the Americas, and US authorities are determined to halt any invasive spread. Keeping them as pets is illegal in many states.

Fact file

- **When hedgehogs meet they bump heads and perform a dance ritual, to establish who is top hedgehog and to protect their foraging trails.**
- **Although they're nocturnal, hedgehogs have poor night vision.**
- **Hedgehogs 'self-anoint' by using their saliva glands to spread a new scent across their prickles, anything from flower scents to less savoury forest smells.**

14. Indian mongoose

Herpestes javanicus

Fearless Rikki-Tikki-Tavi was a *Jungle Book* hero, a tiny slayer of deadly cobras. Such was his reputation that he was taken from Calcutta to Jamaica, to take on the rats that were devastating the sugar-cane plantations. So impressed were the Jamaican planters at the ability of the mongoose to get rid of all sorts of pesky wildlife, they began exporting them to other parts of the world, as far away as Hawaii and even Japan.

Tricky Rikki

Unhappily, it soon became evident that the planters had overlooked one key issue: rats are nocturnal, whereas the mongoose sleeps by night and hunts by day. So instead of wiping out the rodents, as was intended, mongooses enjoyed a good night's sleep then breakfasted on endemic birds, turtles and lizards.

Today this lively and agile animal is blamed for the extinction of birds such as the bar-wing rail (*Nesoclopeus poecilopterus*) on Fiji, and snakes such as the harmless Hispaniola racer (*Alsophis anomalus*). The Jamaican petrel (*Pterodroma caribbaea*) and the pink pigeon (*Columba mayeri*) of Mauritius are also clinging perilously to survival as a result of its depredations. Meanwhile, on the Caribbean island of St. John, mongooses dig up and eat hawksbill turtle (*Eretmochelys imbricata*) eggs. Far away on Japan's Amami island, they are also killing off the Amami rabbit (*Pentalagus furnessi*), a fascinating, primitive species, now reduced to a population of just 5,000 individuals.

With few predators in its various adopted homes, the mongoose is a difficult problem to solve. Some individuals are trapped near wildlife reserves, but there's a feeling on Mauritius that eradicating the mongoose will only work if done side-by-side with rat control. Meanwhile, invasive mongoose populations keep on growing, with each female having litters of up to five young two or three times a year.

Far from assisting farmers with rat control, the Indian mongoose does much more damage to farmers' livelihoods than the rats ever did. It takes chickens, digs up sweet potatoes, then invades papaya and banana groves for dessert. It also spreads serious diseases like leptospirosis and rabies. A recent American study estimates it costs US$50 million a year to undo damage wrought by this not-so-heroic mongoose. Hardly surprising that anyone caught releasing one in Hawaii faces a hefty US$1,000 fine.

Hawaii

Fact file

- **The Indian Mongoose has some biochemical resistance to snake venom.**
- **Ever since *The Jungle Book*, the Indian mongoose has been a popular pet.**
- **This semi-social mammal uses 12 different sounds to express itself.**

15. Stoat

Mustela erminea

New Zealand didn't learn the lesson about the old lady who swallowed a fly. The European stoat, a graceful weasel-like creature, was introduced in the 1880s in an attempt to get rid of another invasive pest, the rabbit. Unfortunately, the stoat began to decimate rare native ground- and treehole-nesting birds. A dramatic downturn in bird numbers was reported by ornithologists within six years of the stoat's arrival.

Kiwi killers

The losses of kiwis to stoat predation have been so great that New Zealand has installed a nationwide egg collection programme. Now, ornithologists raise many kiwi chicks in hatcheries, only releasing them when they are big enough to defend themselves. Authorities have also built huge stoat-proof fences around nature reserves, and there are stoat eradication programmes in national parks and on offshore islands. A stoat can cover about 60km in a matter of weeks. Tolerant of cold conditions, it can cross mountains between temperate rainforests in New Zealand's Southern Alps, and is at home in all habitats. Worst of all, it's a strong swimmer, so the offshore islands that have been traditional safe havens for New Zealand's endemic birds are constantly under threat.

Fact file

The stoat's winter white ermine fur was prized in the Middle Ages and even today is used as trim in ceremonial gowns worn by the Pope and by Britain's Chancellor of the Exchequer.

16. Weasel

Mustela nivalis

If only the weasel ate nothing but mice, the people of New Zealand and of Sao Tome in the Atlantic might be more tolerant of this tiny predator. It was introduced to both countries from the UK to deal with invasive rodents, and is another example of the dangers of introducing one potential pest to catch another.

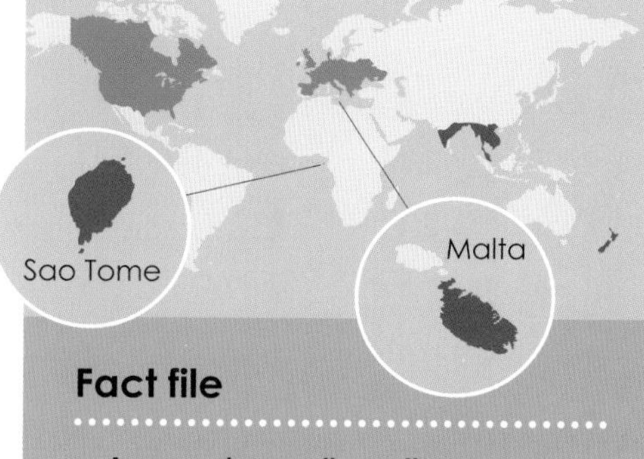

Fact file

- **A weasel can allegedly squeeze through a hole as tiny as a wedding ring.**
- **It needs to kill at least once every 24 hours to survive.**
- **A young weasel, or kit, can hunt from just six weeks old.**

Unfussy eater

Weasels were introduced to Sao Tomé off the coast of west Africa to control mice that were taking the habitat of the endemic and today endangered white-toothed shrew (*Crocidura thomensis*). Unfortunately, the weasels were as keen on shrews as they were on house mice.

In 1885, 67 weasels were released on a peninsula by Lake Wanaka in New Zealand, and soon began preying on native reptiles such as the very endangered Whitaker's skink (*Cyclodina whitakeri*). Naturalists are now so concerned about this lizard that they've begun rounding up the reptiles to rehouse them on weasel-free Mana Island. Weasels also hunt large insects, including New Zealand's endemic tree wetas (*Hemideina*), and will go after native ground-nesting birds, not only destroying eggs, chicks and adults but also taking over their burrows in winter.

17. Domestic cat

Felis catus

The cat has been curling up on the sofa with humans for 4,000 years, since it was first domesticated by the ancient Egyptians. It is intelligent, adaptable and ruthlessly efficient at keeping our homes free of rats and mice. But, though it may be cuddly and appealing, don't be fooled by that innocent expression: this deadly predator is responsible for the deaths of billions of other animals and continues to drive many species toward extinction.

The case against cats

The domestic moggy is descended from the African wildcat (*Felis silvestris*). Although today the two are regarded as separate species, your fluffy pet's hunting patterns and behaviour are almost identical to those of its wild ancestor. The Pilgrim fathers probably bought cats to North America on ships such as the *Mayflower*, and the feisty felines have been behaving like environmental hooligans ever since. The number of cats in the US has leapt from 30 million in the 1970s to at least 70 million kept as pets today. Then there are the feral cats, anywhere between 40 and 60 million, according to the American Bird Conservancy (ABC).

With its devastating hunting skills, the cat is a serious threat to the survival of some of the United States' most vulnerable species. In Florida, the wildlife service estimates cats kill 271 million small mammals and 68 million birds every year, many of them endangered native species. The feral cat, as wild and wily as any non-domestic predator, is a particular threat to hatchling green sea turtles, and rare endemic mammals such as the Lower Keys marsh rabbit (*Sylvilagus palustris*). As few as 100 marsh rabbits survive, and the species is likely to disappear within 20 years if the current mortality rates continue. A study showed

▲ Domestication has done little to quell the feline instinct to capture small animals.

that feral cats were responsible for more than half the marsh rabbit deaths.

Domestic cats are also believed to be responsible for spreading cat-specific diseases such as feline panleukopenia (FPV) and feline leukemia virus (FeLV) to the Florida panther, a distinct form of mountain lion (*Puma concolor*) that is found only in Florida.

Puss in paradise

On islands that are home to rare species, the impact of these felines is nothing short of cat-astrophic. Feral cats have reached the Aldabra Atoll in the Seychelles, which has the largest world population of endemic Aldabra giant tortoises (*Geochelone gigantea*) and is an important breeding ground for the green turtle (*Chelonia mydas*) and hawksbill turtle (*Eretmochelys imbricata*). A study found that baby turtles were the cats' favourite food during hatching season.

On Christmas Island, cats prey on the large flying-fox (*Pteropus melanotus*) and imperial pigeon (*Ducula whartoni*). And in New Zealand, cats saw off the Stephens Island wren (*Xenicus lyalli*). In fact, this now-extinct flightless bird was for many years thought to have been the only species wiped out by a single animal: the lighthouse-keeper's cat. This efficient individual began bringing dead wrens back to the lighthouse soon after arriving on the island in summer 1894, and no more wrens (alive or dead) were seen on the island after spring 1895. Evidence now shows there were other feral cats on Stephens Island, but the story underlines how quickly a species can disappear when cats are on the loose.

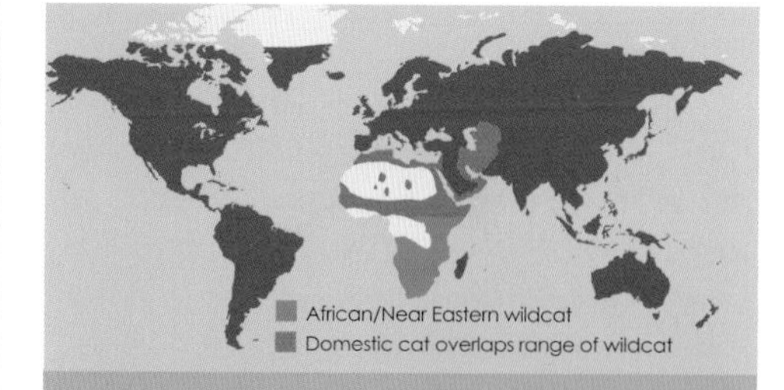

Fact file

- **There are more than 500 million domestic cats in the world, and enthusiasts have developed more than 30 distinct breeds.**
- **A pair of cats and their offspring can found a population of more than 400,000 cats in just seven years.**
- **In Britain, nine million cats kill an estimated 275 million other animals each year – most of them small mammals such as voles and mice.**
- **The best way to stop your cat becoming a wildlife pest is to keep it indoors. Alternatively, fitting its collar with a bell and providing plenty of toys will help curb its hunting.**
- **Sir Isaac Newton is credited with inventing the cat flap.**

18. American mink

Neovison vison

The mink coat that's been a must-have for generations of fashion icons, from Marilyn Monroe to Cindy Crawford, has a much bigger cost than the price tag on the sales racks. The North American mink has escaped from fur farms across Europe, from Ireland to eastern Russia, and had a devastating impact on the natural world.

Please release me

Importing mink for the fur industry began in the 1920s, and there are still many farms across Europe today. The mink's feral population has also been given a massive boost by possibly well-meaning but disastrously ill-informed animal rescue activists who have released them into the wild. As recently as September 2010, some 5,000 minks were set free from a fur farm in south Donegal, Ireland. As one journalist commented, the activists would have done less environmental damage had they tipped a tanker of bleach into the River Finn.

Such a large-scale release may push endangered species over the edge: BirdWatch Ireland was particularly concerned about the red-throated diver (*Gavia stellata*). Donegal is the last Irish breeding ground for this rare bird, with as few as four breeding pairs.

The mink uses streams and rivers to get around, and can survive in harsh arctic winters, in the unforgiving volcanic moonscapes of Iceland and in cities. In Denmark, which produces 40% of the world's farmed pelts, there are minks in most of the canals and harbours around Copenhagen.

Friends and enemies

The losers are native and endemic animals, in particular the near-extinct European mink (*Mustela lutreola*), one of the world's most endangered mammals. A century ago, the European mink was common right across Europe

Farming fur

There are some 6,000 mink farms across Europe, producing two-thirds of the world's mink pelts. Elsewhere, China and America are the biggest exporters of mink fur, with just 300 farms in the US selling US$186 million worth of pelts in a year. Around the world, the industry is now worth US$46.5 million annually. Fur farms are illegal in several EU countries, including the UK.

as far north as Finland. Today it is only found in small pockets of eastern Europe, Spain and France, unable to compete against its larger and more aggressive American cousin for food and habitat. To add insult to injury, American males will mate with European females in the spring before the cuckolded European males are ready for action. No young are born from these American/European alliances, but the females won't mate again that season.

Another native European mammal under threat from invasive mink is the water vole (*Arvicola amphibius*). In Britain, it is easy prey for the mink, which can chase it through water and into its burrows. Water vole numbers across the British Isles plummeted from about eight million in the 1950s to just 220,000 by 2004. Such was the concern for likeable 'Ratty', as the water vole was called in classic children's book *Wind in the Willows*, that Britain gave this little rodent full protection in 2008. It's now being reintroduced to suitable habitats across the country. The vole has another champion in the wild, however. The Eurasian otter (*Lutra lutra*) is one of the few animals not only to fight back but also to prey upon invasive mink. Water vole numbers are recovering wherever there are sturdy native otters defending the neighbourhood.

Outer Hebrides invasion

Ground-nesting birds are easy pickings for the voracious American mink. Introduced to Lewis in Scotland's Outer Hebrides in 1969, it immediately began colonising smaller islands such as Benbecula and the Uists. Minks attacked the nests of vulnerable birds such as arctic terns (*Sterna paradisaea*) and corncrakes (*Crex crex*). BirdLife International estimates 95% of globally threatened birds in the Scottish Islands are at risk from invasive predators, such as the mink.

At the beginning of the 21st century, Scottish conservationists began a massive campaign against the mink, trapping thousands across the islands. After nearly a decade the mink numbers have been drastically reduced in the Hebrides, and bird numbers are starting to recover.

Fact file

- **The favourite food of the American mink in its native territory is the muskrat (*Ondatra zibethicus*).**
- **It has partially webbed feet – making it an excellent swimmer.**
- **The mink prefers to hijack the dens of other animals rather than build its own burrows.**
- **The domestic mink is bred to be larger than a wild mink.**
- **Domesticated, farmed minks have brains 19.6% smaller than native American minks in the wild.**

19. Red fox

Vulpes vulpes

Huntsmen were to blame for the introduction of the red fox to North America and Australia. Missing the thrill of the hunt back home in England, they shipped foxes to their distant estates and helped it become the most widely distributed carnivore on Earth. The fox has long been a respected foe on country estates, being an extraordinarily adaptable animal, omnivorous and able to survive even when its prey is in short supply. One example of its ability to change with the times has been the migration of native British foxes from the countryside to cities in search of easier pickings.

Outfoxed

When the fox arrived in Australia, there were very few foxhunts because farms were so far apart. Nor were there suitable native predators around. These days, the Australian dingo (*Canis lupus dingo*) will take a fox, but it took a long while to recognise this new species as dinner. Foxes now thrive in Australia, except in the most extreme desert and in the tropics.

▲ *A red fox has excellent sight, smell and hearing.*

Brer Fox goes Stateside

Hunters also introduced red foxes to the east coast of North America, in the 18th century. The foreign foxes have spread and thrived, with disastrous consequences for many species of ground-nesting birds and small mammals. Sadly, they also threaten their North American relatives, particularly the delightful San Joaquin kit fox (*Vulpes macrotis mutica*), down to about 7,000 in the wild as red foxes – and humans – steal more of its territory every year.

Victims of the red fox have been unable to adapt to the new threat. The invasive canine is threatening the survival of rare marsupials, such as the brush-tailed rock-wallaby (*Petrogale penicillata*) and the mountain pygmy-possum (*Burramys parvus*). It has also helped spread invasive plants, such as the bitou bush (*Chrysanthemoides monilifera rotundata*), that crowd out native plants and limit the food supply for insects and birds.

Now red foxes have been illegally introduced to Tasmania, one of the most biodiverse large islands on the planet. The authorities have been quick to mount searches, concerned that the island's endemic and native animals are at grave risk. Tasmania is the last Australian refuge for many animals that have been lost to invasive aliens on the mainland, species such as the eastern quoll (*Dasyurus viverrinus*) and the cute bettong or rat kangaroo (*Bettongia gaimardi*). The fox will even take baby Tasmanian devils (*Sarcophilus harrisii*) while their parents are out foraging.

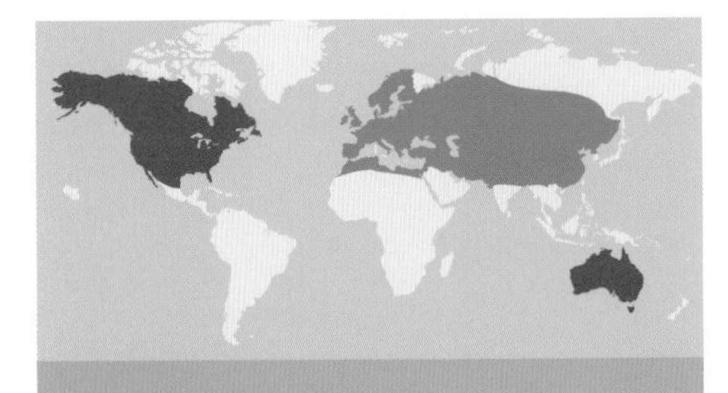

Fact file

- **The red fox is the largest true fox.**
- **It is found on 70 million km² of the planet.**
- **A study in Russia showed red foxes ate up to 300 different animal species.**

20. Crab-eating macaque

Macaca fascicularis

This cheeky monkey is one of life's great opportunists. As its name suggests, it will gorge on crabs, shrimps and frogs in its native mangrove swamps. Yet in the rainforest, families of macaques will mostly browse on fruit, vegetables and seeds, supplementing them with birds' eggs and the occasional chick, given half a chance. Unfortunately, this catholic diet puts the macaque in direct conflict with humans and native wildlife in its adopted homes.

Monkey business

Native to southeast Asia, the crab-eating macaque has been introduced to Indonesia's Irian Jaya, where it is creating serious economic problems for struggling villagers, raiding subsistence crops of cassava and maize. It's also fond of peanuts, mangoes, pineapples and papaya – basically, it likes all the same things its human cousins enjoy. Its activities have had a serious impact on tribespeople in one of the poorest corners of Indonesia. According to the global Invasive Species Specialist Group, farmers can lose US$3,500 in crops to thieving monkeys in a single year.

Across the Indian Ocean in Mauritius, the dextrous crab-eating macaque has learnt to dig up and eat new shoots of sugar cane. This sweet tooth is costing the country an estimated US$3 million every year. And when it's not gorging on sugar, it's taken to swooping on vegetable crops, with sweet potatoes, peppers and pumpkins high on the list of favourites. It is also creating a serious problem for ecologists trying to encourage biodiversity. For a start, the monkey spreads the seeds of introduced and invasive plants. Then it compounds its crimes by taking the eggs and fledglings of endangered birds such as the Mauritius pink pigeon (*Columba mayeri*). Worse, the macaque is believed to be responsible for pushing Mauritius's broad-billed parrot (*Lophopsittacus mauritianus*) over the brink to extinction.

We must share the blame. We shipped these monkeys around the world for the pet and pharmaceutical trade. In its native lands, the crab-eating macaque has plenty of natural predators, such as tigers, sun bears and snakes. However, even in its native environment, its population is liable to soar as rainforests are chopped down and predators killed. The macaque was hunted by forest people, but these indigenous communities are also disappearing as their rainforests are destroyed for oil palm. Ironically, crab-eating macaques are rather partial to oil palm.

Fact file

- The macaque has cheek pouches to carry food and take it back to the young and other members of their family.
- Macaques in Mauritius are sold to the pharmaceutical industry for about US$1,500 each.
- It can live up to 25 years in the wild.

21. Coypu

Myocastor coypus

Fur farmers are ultimately to blame for having exported this hungry herbivore to every continent except Australia and Antarctica. When the coypu fur industry collapsed, the animals were released into the wild to gnaw their way into a whole heap of trouble.

Gnawing pains

The coypu is a semi-aquatic herbivore, burrowing in river or canal banks. It eats 25% of its body weight in river plants every day and is incredibly wasteful, chucking away 90% of a plant to get at the juicy roots. Its activity erodes and undermines riverbanks and levees, and displaces native animals from traditional breeding grounds.

As if that wasn't bad enough, the coypu also spreads disease. It carries a nematode parasite called nutria itch (*Strongyloides myopotami*) that causes a skin disease with symptoms similar to dermatitis in people. It is also a host for other old enemies of mankind, such as tuberculosis and septicaemia.

Unsurprisingly, the coypu – or nutria as it's known in the US – has a price on its head. The rodent causes such widespread destruction to wetlands that Louisiana authorities pay bounty hunters to track them down. The people of Louisiana have good reason to be angry about this South American invader – damage to the levees by coypus considerably worsened flooding during Hurricane Katrina in 2005. The bounty scheme is having some success: the amount of marshland damaged by coypu activity has fallen by 90% since the scheme began in 1999. Yet it's a race against time, because of the coypu's impressive powers of reproduction. The 20 coypu introduced to Louisiana in the 1930s became an estimated 20 million within two decades. Females have two or three litters each year with as many as 13 young in total.

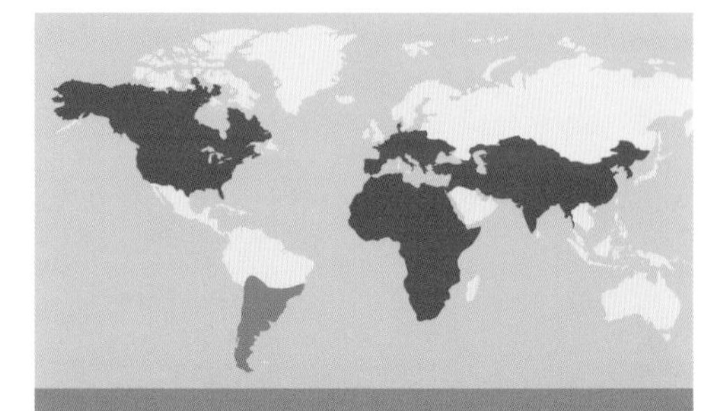

A global struggle

The rest of the world is also fighting back against the coypu invasion. In Kenya, the authorities tried – unsuccessfully – to use introduced pythons to remove coypus from Lake Naivasha. However, in Britain, these invasive rodents have been totally cleared from the waterways of East Anglia. In Germany, Kyrgyzstan and Uzbekistan, farmers are trying to persuade people about the health benefits of eating lean and low-cholesterol coypu meat. Take-up hasn't been great – not many consumers fancy eating what they see as an oversized water rat.

Fact file

- An adult coypu weighs up to 9kg – almost three times as much as a domestic cat.
- It has webbed feet and is a fast swimmer.
- Its rat-like tail helps distinguish the coypu from muskrats and beavers.
- The coypu doesn't like the cold, and suffers from frostbite on its tail.

22. Barbary ground squirrel

Atlantoxerus getulus

In days of old, Fuerteventura in the Canaries was plagued by pirates. They not only invaded the ports but also killed off much of the wildlife. Buccaneers were responsible for the extinction of native seals and possibly the loggerhead turtles – although these are now being reintroduced. Today, another piratical invader is threatening biodiversity on this UN Biosphere Reserve. The tiny Barbary ground squirrel from just across the sea in northwest Africa is upsetting the natural balance of flora and fauna on an island that traditionally had no native land mammals.

Island life

The Barbary ground squirrel was introduced as recently as 1965, from the African Spanish colony of Ifni, probably by the pet trade. It adapted quickly to the arid steppes on Fuerteventura, just 100 or so kilometres from its mainland home. Soon this invasive species had colonised the entire island and today there are about 300,000 of them swarming about Fuerteventura. The squirrel is popular with visitors as it is very cute-looking and by far the most approachable mammal there.

Unfortunately the Barbary ground squirrel disperses the seeds of invasive plants, at a time when ecologists are working hard to protect native species. This spread of plants such as the prickly pear (*Opuntia*) in turn limits the natural food supply for endangered endemic birds such as the Canary Islands stonechat (*Saxicola dacotiae*), once common throughout the Canaries and now only found on Fuerteventura. The squirrel also eats native invertebrates such as snails, which are food for other rare species.

The Barbary ground squirrel is now so widespread on Fuerteventura that scientists believe it may be impossible to eradicate. Instead, they are concentrating their efforts on stopping this adaptable alien moving to the other islands. Transporting the ground squirrel anywhere in the Canaries is strictly illegal.

Fact file

- In northwest Africa, the Barbary ground squirrel mostly feeds on the olive-like fruit and seeds of the argan tree.
- It will migrate when food becomes scarce.
- On Fuerteventura, this adaptable squirrel can now be seen begging tourists for food.

23. Grey squirrel

Sciurus carolinensis

When a London supermarket started advertising grey squirrel cutlets in 2010, there was an outcry from animal lovers. These little bushy-tailed rodents look far too cute to cook, even though Grandma's squirrel pot pie is a favourite dish in the Deep South of the USA. However, if Britain does not find some way of ridding itself of grey squirrels, the monochrome invader could push the indigenous red squirrel (*Sciurus vulgaris*) into extinction.

Going grey

The grey squirrel, from the deciduous forests of North America, was introduced to the UK and Ireland from 1876 until the late 1930s, as an ornamental garden pet. Its population soared in the absence of natural predators, such as skunks, large forest hawks and raccoons. Today, there are about 2.5 million greys ripping through Britain's woodlands, parks and back gardens, and they outnumber the reds by more than 66 to one.

A notorious tree bark-eater, the squirrel is particularly fond of sycamore (*Acer pseudoplantanus*) and beech (*Fagus sylvatica*). In lean food years, it will completely ring-bark trees, killing everything above the nibbled ring line or at the very least making the tree vulnerable to fungal disease. In the year 2000, it was estimated grey squirrels caused £10 million worth of damage to commercial tree plantations. An even bigger concern is the threat to the native British red squirrel. Wherever greys and reds meet, the native squirrels usually disappear within 15 years, outcompeted for food and other resources. Even more lethally, the grey brought with it a squirrel pox virus. The pox is a form of dermatitis that can kill a red squirrel within five days, but is harmless to greys.

Now, the only significant populations of the once widespread red squirrel on the UK mainland are in the Scottish Highlands and isolated pockets of woodland in northern England and Wales. There are also reds on the Isle of Wight, in isolated parts of Northern Ireland, and the Irish Republic.

The grey squirrel has many survival advantages over the red. It's highly adaptable and dextrous (as anybody who has a bird feeder can testify). It has an excellent memory and can find its underground food stores months later. And it's a prolific breeder, having twice as many young as the more reticent reds. On top of all that, a grey lives for about nine years whereas red squirrels fall off their branch after just five or six years.

Lifelines

There is still hope. The Red Alert North England project supports 16 red squirrel reserves, mostly in Northumberland and Cumbria. These are being kept clear of grey squirrels and protected by buffer zones in the hope that reds' populations may grow and recover.

In Scotland, the stakes are even higher. Not only are the Highlands free of greys but this region is home to 75% of Britain's surviving indigenous reds. One key initiative is the See Highland Red project, which has put in place a wide network of volunteers and enlisted public support to protect and map the red squirrels and report any sightings of the greys. Not since the Jacobite uprising have Highlanders mustered such a determined army against an alien invader.

Red squirrels in the Highlands

The red squirrel was widely distributed in Britain until the turn of the 20th century, when the arrival of the greys plus widespread culling saw its numbers start to plummet. The Highlands of Scotland are the last main refuge of Britain's native red squirrel, distinguished from other European reds by its fluffy tail. Today this species is protected and about 120,000 remain in the wild. The best places to see one are within national parks, and it readily visits bird tables. If you do see a red squirrel anywhere in the Highlands, report the sighting on www.redsquirrelsofthehighlands.co.uk.

▲ A gamekeeper's gibbet: grey squirrels in the UK can be legally destroyed.

Fact file

- A grey squirrel can leap more than 6m, and run along the ground at 25km an hour.
- It doesn't hibernate but builds thick winter nests, called dreys, and makes day trips to ground larders for food.
- It can lose its tail sheath and even dislocate tail bones if necessary to escape a predator.
- The goshawk, a large bird of prey once extinct in Britain, is now back and doing its bit to control grey squirrel numbers.

24. Black rat

Rattus rattus

The fortunes of rats and people have been linked throughout human history. The black or ship rat (*Rattus rattus*) originated in India, but thanks to humans is now spread virtually worldwide. It is an environmental bully, directly or indirectly to blame for the extinction of many species of birds, small mammals and reptiles, as well as insects and even plants.

World tour

The black rat has much in common with its human shipmates, being excellent at adapting to new environments. Like sailors of old, it's not keen on swimming, yet recent invasions of New Zealand offshore islands show it's willing to learn if the booty is sufficiently attractive. As an omnivore, it devours just about anything that comes its way.

On small islands, the black rat is particularly destructive. Native island birds have few defences against agile invaders able to shin up branches to raid nests. One such bird was Hawaii's Laysan rail (*Porzana palmeri*). Already wiped out on its native Laysan Island, the rail had been introduced to other Hawaiian islands, but black rats jumped ship from naval vessels during World War II and the bird was extinct by VJ day. Another Hawaiian bird killed off by rats was the Kaua'i 'O'o (*Moho braccatus*), a little honey-eater that fell prey to nest raiding but also to diseases carried by the invasive rodents.

In the Galápagos, the black rat threatens conservation efforts to protect the giant tortoise (*Geochelone nigra*) and marine iguana (*Amblyrhynchus cristatus*), and is blamed for the extinction of eight of the original 12 endemic species of island rodent.

Nor are other species of rats safe from this marauder. On Christmas Island, it took just five years to drive the endemic and numerous Maclear's rat (*Rattus macleari*) into extinction. Recent DNA tests show these large native rats were pushed over the edge by diseases introduced by black rats.

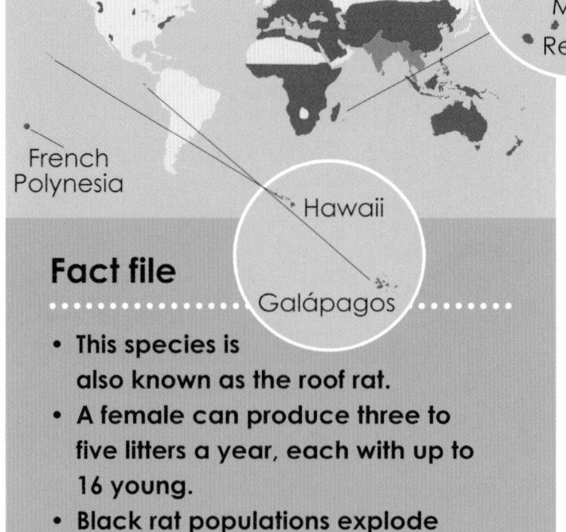

Fact file

- **This species is also known as the roof rat.**
- **A female can produce three to five litters a year, each with up to 16 young.**
- **Black rat populations explode during the rare fruiting of bamboo in Asia, and then move on to devastate crops.**

Rats and humans

Diseases carried by rats have killed more people than all our wars combined. In the Middle Ages, an estimated 200 million people died from the plague (*Yersinia pestis*). There are still up to 3,000 cases reported every year. Such is the continued concern about rats and plague that the US alone spends US$19 billion on rat control every year. Black rats are difficult but not impossible to eradicate. One successful clearance has been on the large Barrow Island in Western Australia.

25. Norway rat

Rattus norvegicus

The Norway or brown rat (*Rattus norvegicus*) is the most successful mammal on earth, after humans, and also one of the most destructive. It invaded Europe in the early 18th century, and by the 1770s had reached the New World via westbound ships. Today it occurs on every continent except Antarctica. It can swim a kilometre across the sea to reach offshore islands, often the last refuge of endangered birds and reptiles.

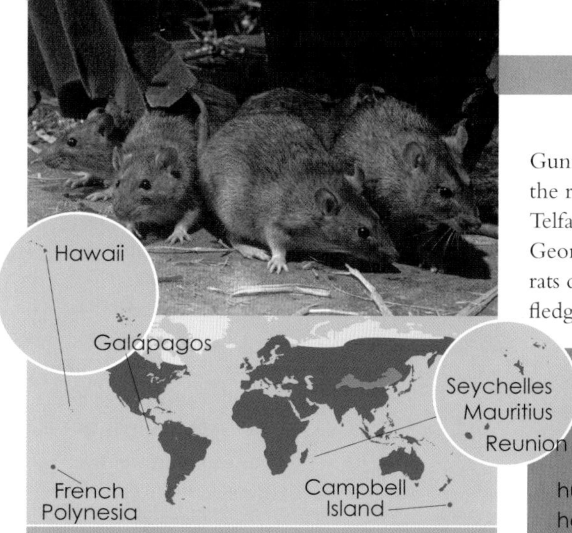

Rampant rodent

In northeastern Brazil, Norway rats destroyed about 3,000 eggs and hatchlings of the critically endangered hawksbill turtle (*Eretmochelys imbricata*) in a single breeding season. On Gunner's Quoin Island off the coast of Mauritius, the rat's arrival led to the fast decline in the Telfairs skink (*Leiolopisma telfairii*). On the South Georgia and the South Sandwich islands, Norway rats devour penguin chicks and take the eggs and fledglings of burrow-nesting small petrels.

Fact file

This rat is not Norwegian at all but hails from northeast China.

Getting rid of rats

Campbell Island in New Zealand's sub-Antarctic region once had a huge population of Norway rats, with a horrendous impact on the population of nesting birds. In the early 2000s, the New Zealand Department of Conservation embarked on the most ambitious rodent eradication project ever attempted. The island is now free of the rats and seabird numbers are recovering.

26. Pacific rat

Rattus exulans

More than 25,000 Henderson petrel chicks – some 95% of all that hatch – are killed every year before they leave the nest. The culprit is the Pacific rat, a dangerous alien introduced to Pacific islands during the mighty Polynesian migrations 1,000 or more years ago. Today, we are still counting the cost of this introduction on once wildlife-rich islands.

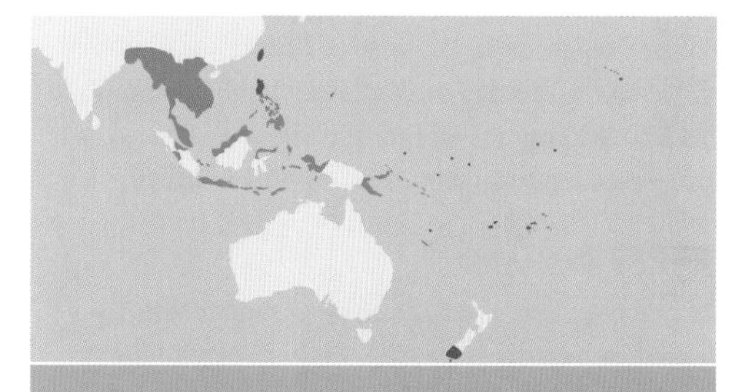

Ratted out

Henderson Island is home to 55 species found nowhere else on Earth. It's the only known breeding ground for the Henderson petrel (*Pterodroma arminjoniana*) and is a vital nesting place for endangered sea turtles. The rats are threatening the entire ecosystem, eating eggs, baby birds, turtle hatchlings and endemic plants. The Royal Society for the Protection of Birds (RSPB) has now begun an eradication campaign and aims to rid Henderson of Pacific rats by late 2013 – the 25th anniversary of the island having been declared a UN World Heritage Site.

The Pacific rat is an efficient predator wherever it is brought to shore. In New Zealand, it attacks unique native insects like wetas, and is to blame for the extinction of some species of gecko and skink.

Fact file

- **This species is the world's third most widespread rat.**
- **A female can have up to six litters of as many as 11 young each year.**
- **It's a poor swimmer, so relies on humans to expand its territory to new islands.**

27. House mouse

Mus musculus

Mice and men have been living together for at least 8,000 years, and we're two of the most successful mammals on Earth. With our help, the house mouse has become the second most widely distributed mammal on the planet. Yet although we live in close proximity, this mouse is one of our oldest enemies, and is a deadly killer preying ruthlessly upon endangered bird species.

Plagues

In 1917, an Australian newspaper reported that farmers in the tiny Victorian township of Lescelles had killed three tons of mice in a single night. By the end of that plague season, they'd caught a staggering 32 million mice. Such plagues of mice now hit Australia every three or four years and occur when rainy seasons follow a long period of drought. Mice do an estimated A$36 million worth of damage to 500,000ha in a season. At that time, there will be 3,000 or more mice for every hectare outdoors and 52,000 squealing rodents overrunning each hectare of grain stores. There are regular mouse plagues in northwestern China and occasionally in California, but nothing in comparison with the Australian scourge.

Getting rid of the house mouse long term is extremely difficult because of its astounding birth rate. The females start giving birth to young when they're just five weeks old and a breeding pair can produce up to 500 mice in 21 weeks. The female will mate again the day after giving birth. Trapping and poisoning are effective forms of control, but poisons threaten farm animals and wildlife.

Killer mice

Mobs of giant killer mice are slaughtering up to a million seabirds every year on Gough Island in the South Atlantic. There are no rats on the island so marooned mice have grown abnormally large and their population has exploded to about 700,000. Gangs of up to ten mice swarm on the nests and literally eat the chicks alive. This is disastrous for the endangered Tristan albatross (*Diomedea dabbenena*) that only breeds on this isolated island. The mice also devour great shearwater (*Puffinus gravis*) and Atlantic petrel (*Pterodroma incerta*) chicks. Another species they have pushed to the brink is the critically endangered and endemic Gough bunting (*Rowettia goughensis*), which was common across the island back in the 1920s.

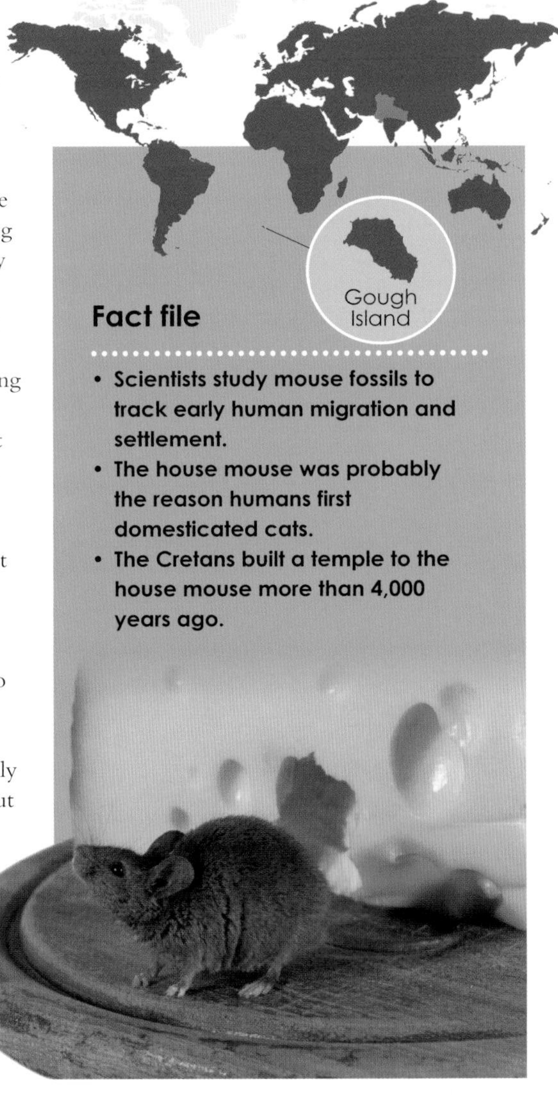

Fact file

- **Scientists study mouse fossils to track early human migration and settlement.**
- **The house mouse was probably the reason humans first domesticated cats.**
- **The Cretans built a temple to the house mouse more than 4,000 years ago.**

28. Edible dormouse

Glis glis

Feeling a bit peckish but no time for a sit-down meal? Then how about dropping into a snack bar for a fried dormouse? That would have been the case in ancient Rome where this pretty dormouse was frequently served up as a tasty treat. Dormice destined for this fate were raised in little terra cotta containers called gliraria. It was a bit like keeping a hamster in a cage. Even today, edible dormice are considered a delicacy in Slovenia.

Chewing it over

This succulent rodent is native across continental Europe and across into Iran, Turkey and some Mediterranean islands. Unfortunately it has now also arrived in Britain, where is has become something of a problem, threatening ancient woodland in southern England. It strips the bark off mature trees and is particularly fond of larch. It also attacks fruit crops. In winter, some dormice may move indoors, getting into roof spaces and nibbling through electrical wiring and insulation.

Slow spread

Unlike many other invasive mammals, the edible dormouse has not been too quick to exploit its new territory. There are several reasons for this. It only breeds once a year, so its population doesn't build that quickly. It is also very dependent on tree-rich habitats – it would have much preferred the England of two centuries ago, with its more continuous tree cover. And, because the dormouse comes from nearby continental Europe, many of the predators that hunt it there also occur in Britain. Despite all that, it is slowly but surely building in numbers.

Don't blame the dormouse, however. The fault lies with Lionel Walter Rothschild, who imported dormice for his private animal collection at Tring in 1902. These escaped and now there are at least 10,000 of them in the Chiltern hills. There have also been reports of dormice chomping their way across Dorset, Hampshire and into Essex. Although they don't breed as fast as mice, these rodents live up to 12 years, and scientists believe their population will expand as the climate heats up.

Fact file

- The edible dormouse looks a bit like another introduced UK mammal, the grey squirrel, but is smaller and daintier.
- It hibernates for up to six months of the year.
- Despite being a poor jumper, it spends most of its waking hours high in the treetops.

29. European rabbit

Oryctolagus cuniculus

The European rabbit has hopped into a whole lot of trouble on every continent except Asia. For such an apparently harmless herbivore, it has caused disproportionate damage to ecosystems and agriculture, and ultimately may have brought about the extinction of many other species. It is a pest in many parts of the world, but nowhere more so than in Australia where it is blamed in part for the extinction of an eighth of all endemic mammal species.

Menace to marsupials

One such species is the lesser bilby (*Macrotis leucura*) last seen back in the 1950s. Now that species's close relative, the greater bilby (*Macrotis lagotis*) or rabbit-eared bandicoot, is under serious threat from food and habitat loss caused by rabbits. Competition from rabbits is also threatening the tiny brush-tailed bettong (*Bettongia penicillata*) or woylie. The territory occupied by this little animal has fallen from 60% down to just tiny pockets of Australia. All of these marsupials look somewhat rabbit-like, and use a similar environmental niche to that used by rabbits in their native lands.

Rabbits sailed to Australia with the First Fleet in 1778, but only became established on Tasmania in the early 19th century. In 1859, Victorian farmer Thomas Austin released 24 rabbits on to his land, intending to use them for meat. By 1910 their descendants had colonised all but the very far north of the continent. Even the 1,700km rabbit-proof fence built between 1901 and 1907 to keep rabbits out of Western Australia failed to halt the advance. Less than a century later there were an estimated 600 million rabbits in Australia.

Rabbits compete – very successfully – with endemic animals for food, which becomes a particular issue during droughts or after bush fires when resources are scarce. They also cause erosion by eating vegetation and burrowing, which reduces ground cover for small mammals.

Rabbit control

The deadly rabbit myxoma virus (which causes myxomatosis) was introduced to Australia in 1950, and to South America three years later. Rabbits initially succumbed to the disease but soon built up immunity. However, this immunity decreases with generations and there are still isolated outbreaks.

They can strip the countryside bare. In the Norfolk Island group, rabbits turned Philip Island to bare earth. Happily, they've since been eradicated, and a programme of reforestation is under way.

Another rabbit-related problem is that the explosion in their numbers supports population growth for introduced predators such as the red fox. Their presence led to the disastrous introduction of the stoat to New Zealand.

Invading a new continent

The European rabbit is also considered a dangerous invader across Argentina and Chile in South America. In 1936, two pairs of rabbits were let loose on the Chilean side of Tierra del Fuego. By the 1950s there were 30 rabbits per hectare and more than a million hectares of Chile covered – an estimated 30 million rabbits! As well as agricultural damage, rabbits cause huge losses in Chile's commercial pine forests by nibbling the young seedlings.

The problem in South America has been compounded by a shortage of predators. While Spanish and Portuguese rabbits are everyday prey for foxes and hawks, South American animals were initially confused by the zig-zag escape path of fleeing rabbits. However, in recent years, birds such as the black-chested eagle (*Geranoaetus melanoleucus*) and the Magellanic horned owl (*Bubo magellanicus*) have learnt to catch rabbits. The chase is on.

Fact file

- A female rabbit can have up to 40 young each year.
- The rabbit is classified as near threatened in its native Portugal and Spain.
- It was first exported in about 1,000 BC by the Phoenician traders, then domesticated by the Romans.
- Rabbits were released on to islands as food for marooned sailors.
- The Dutch shipped rabbits to South Africa as early as 1654.

30. Sika deer

Cervus nippon

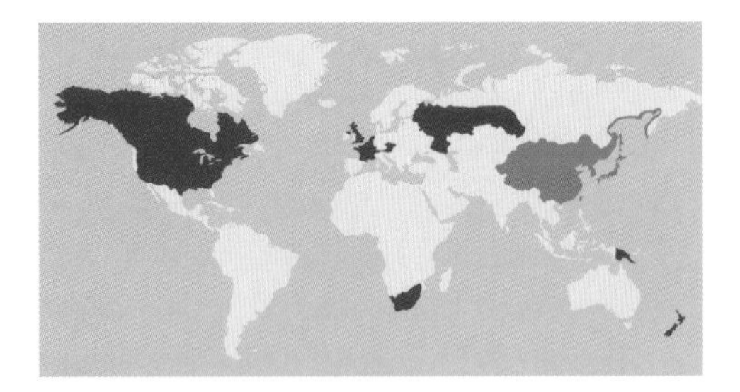

The sika deer puts Houdini to shame. It was brought to the British Isles in the 19th century, and has wriggled free to become a serious threat to the survival of native species. One such Great Escape was from Brownsea Island in Poole Harbour. At the first low tide, the more resourceful members of the herd paddled and swam their way to the mainland. Then King Edward V11 gave a pair of sikas to Baron Montagu of Beaulieu. They soon leapt the fence into Sowley Wood, and their descendents now roam in large herds in the New Forest.

Fortune sikas

Today, the sika is spread across England and Scotland. It's also common in Ireland, thanks to aristocratic huntsmen like Lord Kenmare. He introduced the species into the woods of County Kerry back in the 1860s, and by the 1940s there were an estimated 1,000 sikas in the district.

The damage done by this graceful escapee to forests, heathland and native wildlife is extreme. Sika deer kill large mature trees by peeling away the bark. They also gouge the trees with their antlers during the autumn rut. Outside the forest, sikas damage heathland by stripping away the ground cover plants, leaving the earth bare and vulnerable to erosion. And they're steadily expanding. Prime habitat can hold up to 45 sikas per hectare, with the population spreading as young males are forced to wander elsewhere in search of fresh territory.

Sika males also woo the native red deer. The species are interbreeding and, worryingly, some deer experts believe there may one day be no such animal as a pure red deer surviving in its native territories. As if that wasn't enough to make the native sika a public enemy, it also carries diseases that are transferred to native animals, including the red deer but also bison and roe deer further afield across continental Europe.

Even on its home ground, the sika is becoming an environmental problem. It's a sacred animal in Japan, where a Shinto god is believed to have ridden a sika into the great Nara. Unfortunately, the deer population has exploded since their main predators, wolves, were hunted to extinction. Now they are stripping the forests of Japan faster than the trees can regenerate.

Fact file

- **This deer spends the day in woodland, for camouflage, but comes out into the open to feed after dark.**
- **Males whistle so loudly during mating season that you can hear them half a mile away.**
- **Japan has the largest numbers of sika deer in the world.**
- **'Sika' is the Japanese word for 'deer'.**

31. Red deer

Cervus elaphus

The stag immortalised by Edwin Landseer's painting *Monarch of the Glen* may be a stirring sight, but these days the red deer is anything but noble. The fault, of course, lies with people, who introduced the red deer for trophy hunting into South America, New Zealand and Australia in the late 19th and early 20th centuries. We also wiped out all the large predators that kept deer numbers in check in Britain.

Deforestation

A few red deer set free in the foothills of the Andes now cover at least 50,000km^2. The red deer has chomped its way through the undergrowth in some of the Earth's most biodiverse temperate forests. One of the worst-hit ecosystems is Chile and Argentina's Valdivian temperate forest, along the west coast. It's the second largest of only five temperate rainforest ecosystems remaining on Earth, and is densely forested with Antarctic beech (*Nothofagus antarctica*) and plants dating back to the Gondwana supercontinent. Valdivian is home to a large number of reptiles and endemic birds, such as the Chilean tinamou (*Nothoprocta perdicaria*).

Unfortunately, the red deer eats the seedlings and saplings of Valdivian's broadleaved trees, meaning the forest fails to regenerate. It also competes with native huemul deer (*Hippocamelus bisulcus*). Naturalists working in Argentina's Nahuel Huapi National Park discovered that the huemul's favourite food, lenga beech (*Nothofagus pumilio*), was also top of the menu for the invasive red deer.

In Britain, the native red deer was driven almost to extinction by hunting, but the Victorians bolstered its numbers with introductions. Today, with no wolves or lynxes to control their numbers, red deer seriously threaten forest regeneration in the British uplands.

The big guns

Enthusiastic New Zealand hunters introduced red deer from Scottish and English estates, with more than 200 separate releases between 1851 and 1914. Ironically, these deer were protected so the herds could expand – and so they did, disastrously and way beyond expectations. Soon large herds were not only destroying farmland but also removing the undergrowth from forests that held the soil in place. Red-faced officials were forced to organise mass culls. A Deer Menace Conference was held in the 1930s and a bounty put on the tail of the red deer. By the 1950s, about 60,000 deer were being shot each year. Yet decades later, the invasive red deer is still an environmental menace.

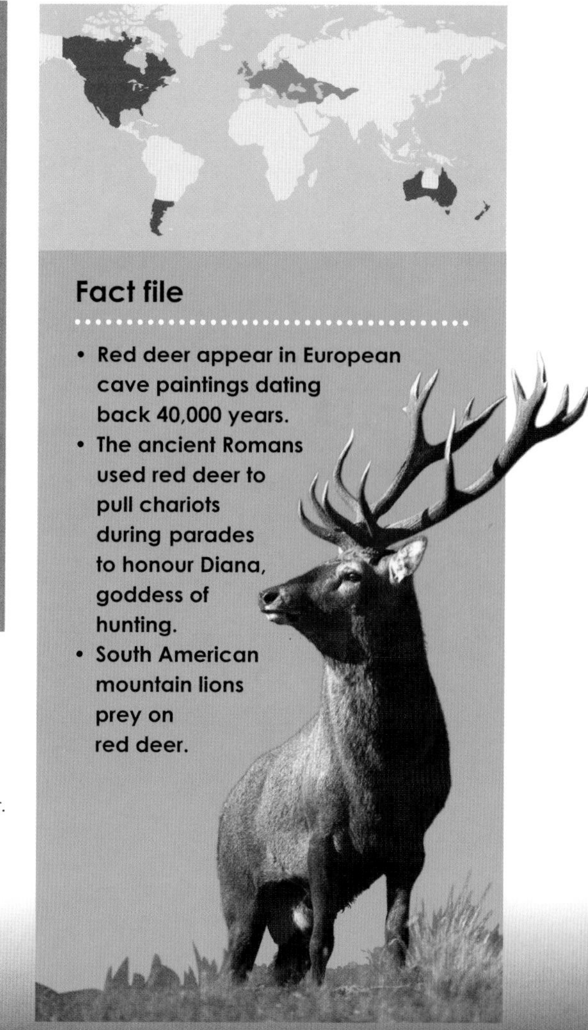

Fact file

- Red deer appear in European cave paintings dating back 40,000 years.
- The ancient Romans used red deer to pull chariots during parades to honour Diana, goddess of hunting.
- South American mountain lions prey on red deer.

32. Feral camel

Camelus dromedarius

Australia wouldn't have a trans-continental railway if it wasn't for the Arabian camel, or dromedary. This beast of burden was imported into Australia by the thousand from 1840 and 1907 to help lay tracks and telegraph lines. Able to carry 800kg, camels were also used to ship supplies into the new mining colonies, and provided transport across the harsh Outback.

Why Aussies have the hump

However, the camel was soon overtaken by the railway it helped to build and by the arrival of motor vehicles. Set free in the bush, it adapted very nicely to its new home and spread fast, soon colonising the wild spaces of the Northern Territory, Queensland and South Australia. Today there are about a million camels in Australia.

Feral camels cause huge damage to wildlife and endemic plants. The worst problem in this drought-prone country, however, is their powerful thirst, which can deprive native species of water. Camels also foul waterholes and eat the waterside plants that provide food and shelter to endangered marsupials. They have been blamed for the disappearance of plants like the wild quandong (*Santalum acuminatum*) and native apricot (*Pittosporum augustifolium*). They also compete against farm animals for water and plants, doing an estimated A$10 million worth of damage every year.

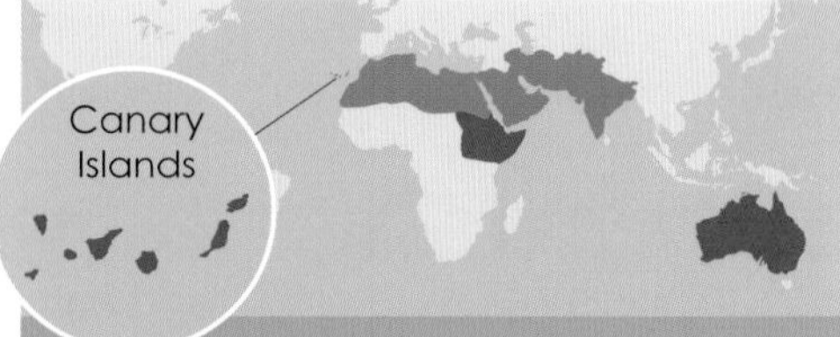

Fact file

- Humps store fat, not water.
- Camels were first domesticated 6,000 years ago.
- They can survive 17 days without water.

---

33. Donkey

Equus africanus asinus

Eeyore – specifically *asinus*, the world's most common breed of donkey – has a lot to answer for. A mature donkey eats as much as 2,766kg (6,000lb) of fodder in a year and the species breeds rapidly. It's wonderful as a domestic beast of burden, but when it breaks out into the wild it can proliferate and do real damage. In California's Mojave National Preserve, the population of feral donkeys is estimated to be growing by 25% every year.

Hot stuff

Having started life in the deserts of Arabia, the donkey does well in warm dry places like the Mojave Desert. It damages the soil and plants with its hard hooves, and takes food from under the noses of the native desert tortoises – already on the Red List of endangered species.

In Australia, donkeys are on the government's most wanted list, for eating native plants and spreading invasive weeds in their dung. They are a particular danger to wildlife during droughts, when they inconsiderately foul the waterholes that native animals rely on for survival.

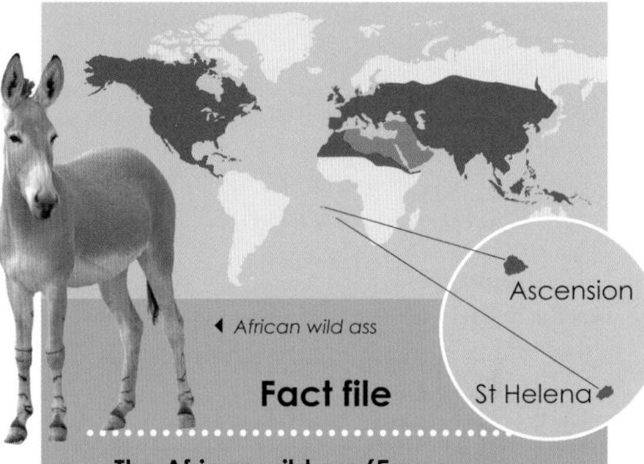

◂ *African wild ass*

Fact file

- The African wild ass (*Equus africanus*), from which the domestic donkey is descended, is now a critically endangered species.
- Donkeys have been domesticated for about 4,000 years.

34. Feral pig

Sus scrofa

If it had only stuck to truffles, the feral pig wouldn't be up to its little piggy eyes in trouble. We began domesticating the wild boar back around 4,900 BC, and the modern domestic pig is one of our most useful animals. Unfortunately, when pigs escape and establish large, free-roaming bands they can wreak havoc. Their habit of preying on endangered species such as baby land tortoises or newborn fawns has put them among the worst invasive aliens on our planet.

Swine of the times

On Santiago island in the Galápagos, marooned pigs became a serious danger to giant tortoises. They were eating tortoise eggs and hatchlings, then competing for food against the adults. The feral pigs were also raiding sea turtle nests and killing albatross, shag and booby chicks. Such was the damage that the Galápagos National Park authorities decided the porkers were for the chop. In the early 1970s they began a 30-year project that eradicated 18,000 pigs, the largest clearance of its type ever attempted. By the early 2000s, Santiago was swine-free.

Feral pigs were often deliberately marooned like Ben Gunn on isolated islands, as rations for passing sailing ships or shipwreck survivors. They generally flourished, being omnivores and able to eat whatever food was available. When they weren't scoffing eggs and small animals, they'd root up native plants, throwing the whole ecosystem out of kilter and leaving bare patches that are vulnerable to invasive weeds. In Hawaii, pigs have done huge damage to the koa-ohia forests by debarking and felling native tree ferns.

Pigs are intelligent and adaptable, changing their behaviour to suit the local conditions. In India, they've become nocturnal, raiding crops after dark so they won't be seen. They're as much at home in the beech forests of America's Smoky Mountains as they are in the marshes of the Deep South or 3,000m-high rainforests in central New Guinea.

Pigs Down Under

Pigs were shipped to Australia through the 19th century, and today there are as many as 23 million feral pigs rampaging across the countryside. They gather around billabongs during droughts in mobs of up to 100 porkers, wrecking important waterside breeding grounds for endangered amphibians like the white-bellied frog (*Geocrinia alba*).

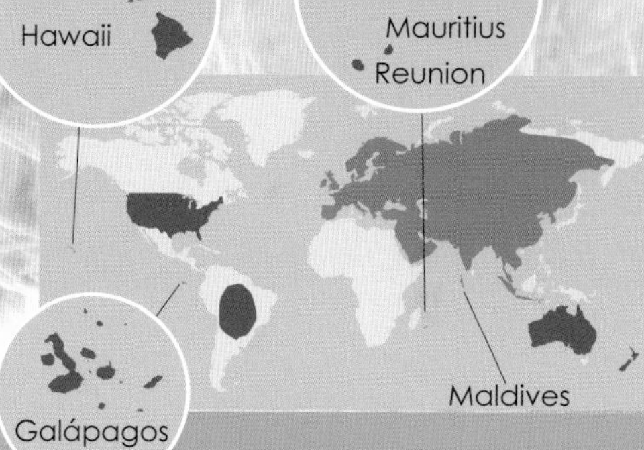

Fact file

- **A wild boar can weigh more than 200kg.**
- **Captain Cook traded tools for pigs during his South Seas expeditions.**
- **In Australia, tagged pigs have travelled 20km in two days to find water.**
- **Pigs don't have sweat glands, so roll in mud to keep cool.**

35. Sheep

Ovis aries

There are about a billion domesticated sheep on Earth. In New Zealand they outnumber people 12 to one. The familiar domestic sheep of today is descended from wild Asiatic (*Ovis orientalis*) and European (*O. musimon*) mouflons and we started farming it 11,000 or so years ago. It is an extremely useful animal, giving us wool, meat and milk. Yet turned feral, this grass-gobbling machine becomes a dangerous alien invader.

Fact file

- There are more than 200 breeds of domestic sheep, varying in coat length, colour and texture, and horn size and shape.
- About 8% of sheep are believed to be homosexual.
- Sheep aren't stupid – studies show they're as smart as cows and almost as clever as pigs, the Einsteins of the farmyard.

Raising the baa

Sheep are incredibly good at adapting to new environments and will survive not only in grassland but in desert and dense forests. When domestic sheep turn feral, they become far less choosy about what they eat, chewing away at native plants that are an important food source for native insects and birds.

Their worst impact is on islands such as California's Santa Cruz, where plants have evolved without herbivores so don't have natural defences like prickles or poison. When these plants disappear, so do the nutrients in the soil and seedlings have trouble growing. So there are fewer trees and the result is a virtual desert – which is disastrous for the animals that depend on the plants too.

Sheep also compete for scarce food supplies with endangered mammals. In Tibet, the critically endangered Przewalski's gazelle is being starved out by greedy sheep.

Unique ferals

Some of the world's populations of feral sheep have been isolated from other sheep for so long that they have developed their own distinctive appearance. In the St Kilda islands, off the west coast of Scotland, there are two unique forms of sheep, the Soay and Boreray, both descended from ferals. The Santa Cruz sheep formerly occupied Santa Cruz Island in the Channel Islands of California, but all individuals were removed to protect the island habitat. These and other distinct island feral forms are now regarded as 'rare breeds' and enthusiasts are working hard to preserve them.

▾ Sheep introduced to the Falkland Islands compete for habitat with Magellanic penguins.

36. Goat

Capra aegagrus hircus

A goat has its uses. More people drink goat milk than that of any other animal, and it is the most commonly farmed animal on Earth. Unfortunately, it is also responsible for a huge amount of ecological damage to fragile environments around the world. There are goats on every continent except Antarctica, about 4.5 billion of them in total at the last count. The goat builds its population quickly because it breeds so fast: a nanny starts having kids at ten months and has offspring twice a year.

Getting your goat

This tough, hardy and adaptable herbivore thrives in environments that would test the resources of many other creatures. Agile and sure-footed, with a cast-iron digestive system, it is undaunted by tricky terrain and a diet of tough and dry vegetation. Naturally, it takes even better to more hospitable habitats.

Goats were marooned by whalers on the Atlantic island of St Helena in the 16th century. They soon chomped their way through half the endemic plant species. Goats sailed to Australia with the First Fleet and by the 1990s there were at least 2.6 million running wild, causing massive damage to indigenous plants and competing with wildlife for food. San Clemente Island in California is only 14,800ha, but is home to about 29,000 goats.

North Isabela Island in the Galápagos was free of goats until the 1970s. A few were released accidentally and within two decades there were about 150,000 on an island that is home to the world's largest population of giant tortoises. The consequences of this invasion would have been even more serious had prompt action not been taken. The National Park Service began a massive programme to get rid of the goats and the island has recently been declared goat-free.

▼ *Feral goats beside a dingo fence in Australia's outback*

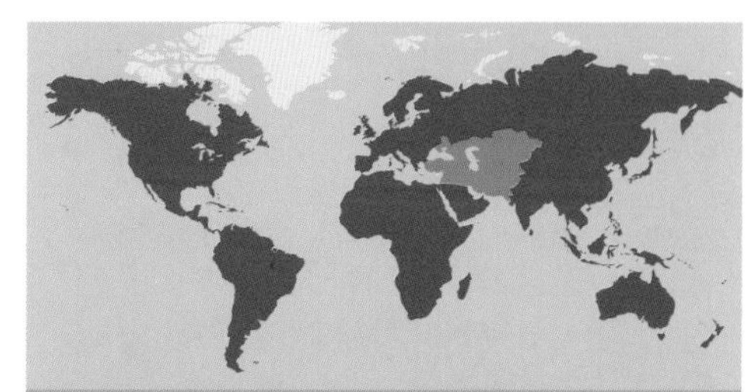

Fact file

- The domestic goat belongs to the cattle family (Bovidae) and is thought to be descended from the wild goat *Capra aegagrus*.
- There are more than 210 breeds of goat.
- Conservation officers use so-called Judas goats to lure feral goats to areas where they can be shot.
- The Egyptian pharaoh Cephranes liked goats so much that he had 2,234 buried in his tomb.

37. Brushtail possum

Trichosurus vulpecula

On New Zealand there are at least seven Australian brushtail possums for every human, and this marsupial is the most despised of all the country's introduced animal pests. Every night these furry invaders munch through 9,000 tonnes of leaves, flowers and fruit. They destroy the native rainforests by concentrating on the new growth. They will return to the same tree night after night until it dies and in some parts of New Zealand have destroyed whole canopies of species such as the native rata (*Metrosideros robusta*) and Kowhai (*Sophora microphylla*). Today possums occupy nearly 99% of the New Zealand mainland.

Fur better, fur worse

The possum was introduced to New Zealand around the 1830s to create an instant fur industry. Some 20 million were trapped in a single year in the height of the fur trade, yet this made little impact on the introduced possum population. In its native Australia, its numbers are kept under natural control by predators such as dingoes and by forest fires. It is also not so well fed. New Zealand trees have few natural defences so were fair game for hungry marsupials.

The degradation of the forest has in turn disastrously restricted the food supply for New Zealand's endemic birds. The possum goes after the high-energy fruit and berries needed by species such as the bellbird (*Anthornis melanura*) during their breeding season, and also takes the eggs of ground-nesting birds. It also attacks New Zealand's two endangered bat species, the country's only native land mammals, and eats the flowers of the native wood rose (*Dactylanthus taylorii*), a nectar source for bats. Endemic snails and wetas are on the menu too.

Naturally adaptable, possums live in most habitats, although they prefer hardwood mixed forests and thrive in mixed farming and forestry areas. As well as being the fluffy kiss of death to wildlife, they spread bovine tuberculosis to farm animals in New Zealand.

Moa's Ark

New Zealand was nicknamed 'Moa's Ark' by botanist David Bellamy, because of its unique plants and animals – including the giant, flightless moa – that evolved for millions of years without mammalian predators. When the Polynesians arrived in about the 8th century, they brought the first alien mammals to this undisturbed environment. Europeans followed with a whole raft of other predators that cut a swathe through the population of native birds, bats, amphibians, snails and insects. These endemic animals were not equipped by evolution to deal with the onslaught, and many species became extinct in an alarmingly short time.

Waging war

New Zealand declared war on the alien marsupial in the 1940s, when keeping and releasing brushtail possums was declared illegal. Since that time the government has spent multi-millions of dollars trying to eradicate the pest. A bounty scheme in the 1960s failed when police discovered fraudulent hunters were releasing additional possums into the forests, then recatching them to claim the reward! Other schemes have worked better. From an estimated 70 million back in the 1980s, there are now believed to be around 30 million possums.

Fact file

- The brushtail possum is an expert climber, with a prehensile tail.
- A possum can make 22 different sounds, from squeaks and clicks to hisses.
- The scientific name *Trichosurus vulpecula* comes from the Greek, meaning 'furry-tailed little fox'
- Tiritiri Matangi, an island reserve, was cleared of its possums and is now home to rare birds such as the little spotted kiwi.
- The possum has not yet conquered the southwestern Fiordland forests, and conservationists are desperate to keep this pristine forested region possum-free.

Today, New Zealand's Department of Conservation is trying to reverse the fortunes of the country's surviving wildlife by clearing invasive predators from small offshore islands, replanting indigenous forest and reintroducing endangered species. The conservationists now have 13.3 million hectares under possum control schemes, and their efforts are reaping rewards. When possums were cleared from one patch of native forest, the population of rare New Zealand robins (*Petroica australis*) rose by a third.

Meanwhile in the capital, Wellington, conservationists have built an 8.6km predator-proof fence around a former city park to create the first 'mainland island' reserve. Karori wildlife reserve and wetland covers 225ha and is home to giant wetas, tuataras and little spotted kiwis (*Apteryx owenii*) – at last safe on the mainland from possums and their ilk.

3 UNDERWATER ALIENS

Just when you thought it was safe to go back in the water… Great battles are being waged beneath the surface of our oceans, lakes and rivers as marine and freshwater species struggle to withstand the onslaught of exotic invaders.

Lionfish devastate native fish populations on the reefs to which they have been introduced.

Aquatic animals have always floated merrily downstream, or bobbed along in ocean currents, to colonise new territory. They drift with the tide, and are thrown up on shores hundreds of kilometres away by fierce storms. Yet their travel ambitions are naturally limited by wind direction, water currents and bigger animals with sharp teeth. As a result, unique and isolated ecosystems have evolved, and – just like ecosystems on land – these are vulnerable to invasion by alien species.

Water ways

Since humans began exploring the world, aquatic invaders have gained new ways to move from one ocean to the next, and have crossed the globe to terrorise the natives of distant lakes and river systems. In the 21st century, aquatic invaders are believed to pose the greatest single threat to biodiversity in coastal waters worldwide. They cause numerous problems, from choking the mudflat habitats of waterfowl to infecting drinking water reservoirs and spreading exotic diseases to seafood farms.

Census of marine life

The oceans of the world are home to more than a million species, and we've only identified about a quarter of them. That's the verdict from a decade-long study by scientists from 80 countries around the world. Remarkably little was known about the wildlife of our oceans at the turn of the 21st century. This global project, called the Census of Marine Life, set out to fill that gap in our knowledge by counting and mapping the creatures of the deep. The idea is to give scientists a benchmark against which they can measure changes in marine biodiversity caused by warming seas, pollution and the inevitable attack of alien invaders.

▼ *A sea bass farm off Ugljan Island, Croatia: fish farming in some areas is associated with the introduction of disease to native marine species*

▼ *Goldfish and other aquarium species can cause problems when released into aquatic ecosystems*

These aliens owe their success to shipping, fish farms and fishbowls. From the 18th century onwards, aggressive underwater species have been shipped en masse around the world, with a disastrous impact on freshwater and marine ecosystems. Not all introduced species survive and thrive in their new territory but, since the mid-20th century, rising sea temperatures have been helping tropical marine animals colonise new parts of the world that would have been too chilly for them a century ago. Invasion usually takes several years, which means an invasive species is often well established in its new territory by the time it is identified as a significant problem.

Cruising for a bruising

Many of the world's most pernicious aquatic invaders have stowed away by clinging to the hulls of cargo ships. Others are carried in what is known as ballast water. About 80% of the world's freight is transported by sea, and ships empty or take on water in tanks in the hull. The weight of this ballast water keeps the vessel balanced when cargo is loaded or unloaded. It's not unusual for a cargo vessel to carry 70 million litres, and some ten billion tonnes of ballast water are ferried around the world every year. It's an environmental tidal wave. When the ships dock, the water is flushed out and so are all the tiny passengers: millions of planktonic animals and their larvae.

One such stowaway in ballast water was the zebra mussel from eastern Europe, now causing serious eco-problems in North America's Great Lakes (see page 64). Barnacles (see page 67) also love to hitch a ride in ballast water, as do invasive marine plants and algae, such as Japanese kelp (see page 146). Most of these invasive species cause problems around ports or in lakes, and many can pose a serious threat to human health. A cholera epidemic in Peru during the 1990s that killed 9,600 people was blamed on seafood contaminated by ballast water sucked up as far away as Asia.

Governments around the world are now struggling to find ways to solve the ballast water problem. Having ships empty their bilges further out to sea goes some way to easing the problem. Sterilising ship holds and using chemicals to kill off the nasties are probably more effective solutions. However, the shipping industry is not the only means of oceanic travel for invasive species. Every year, we throw masses of plastic rubbish into the seas. This provides rafts for intrepid seafarers, from aquatic bugs to molluscs and even plants.

The problem with fish suppers

Another reason alien species are spreading around the world is our appetite for live seafood. Not only do these creatures escape from restaurant tanks (and who can blame the fish?) but the seawater used to ship sea creatures is often emptied into the bay and may contain the eggs of unwanted species. Many other aquatic pests have escaped from fish farms or been deliberately released into lakes and rivers as a source of food or as sport fish. One of the worst is the Nile perch (see page 77), a voracious predator that has caused the extinction of hundreds of cichlid fish species in Lake Victoria.

▼ *Lake Victoria cichlid*

38. Zebra mussel

Dreissena polymorpha

A new Cold War is under way, and this time the Russians are winning. Zebra mussels from Russia's southeastern lakes and the Black Sea have invaded North America, with devastating results for the local wildlife. This pesky mollusc was carried to North America in the ballast water of cargo ships in the late 1980s, and was first identified in Lake St Clair. Now it has spread across the Great Lakes, and beyond to streams and rivers as far afield as New England and California. Nor is it an exclusively North American problem: the zebra mussel has been also terrorising the local molluscs in the British Isles, Sweden, Spain and Italy's Lake Garda.

Overwhelming numbers

It's hardly surprising that the zebra mussel soon took hold in foreign waterways, when you consider the fact that a single female can produce a million eggs every year. As many as 10,000 young mussels may attach themselves to a single North American freshwater mussel. They'll engulf a clam so that it cannot open to feed, and even attach themselves to turtles and crayfish.

The zebra mussel has a remarkable filtration system and soon makes lake water look crystal clear. However, this cleaning process filters out the algae that form a food supply for other native animals. Ironically, there have been occasional beneficiaries – numbers of yellow perch in Lake St. Clair and smallmouth bass in Lake Eerie have risen since the arrival of the zebra mussel. But don't be fooled. Not for nothing is the zebra mussel labelled as the most aggressive freshwater invader on Earth. This stripy invader is blamed for bringing avian botulism to the Great Lakes – a devastating disease, which has killed thousands of waterbirds, including loons, ring-billed gulls and grebes, since the late 1990s.

It is also an expensive drain on North American taxpayers. Clearing huge colonies of zebra mussels from pipes around water treatment and hydro-electric plants costs an estimated US$500 million every year.

Fact file

- Boat owners can help prevent the further spread of zebra mussels by thoroughly cleaning their vessels and disinfecting the bilges.
- A zebra mussel can survive for weeks out of water.
- It can live for up to five years.

39. Crown-of-thorns starfish

Acanthaster planci

These aliens swoop like an invading army, up to 200 at a time, turning a pristine coral reef into a skeletal underwater desert. The crown-of-thorns starfish literally sucks the life out of the reef by attaching itself to the coral and dining on the living polyps that form these fabulous structures. It is able to squeeze its stomach completely outside its body, and then secretes enzymes that turn the coral to mush. In just one year, a single starfish can destroy 10m² of reef.

Coral catastrophe

This strikingly spiny starfish is responsible for the Swiss cheese effect noticeable on threatened reefs, from Mauritius in the Indian Ocean to the Great Barrier Reef in Australia, and even the Red Sea (named by legendary underwater explorer Jacques Cousteau as one of the ten great marine sites in the world). In 2009 there was also a massive invasion on the coral within the Philippines' World Heritage Tubbataha Reefs Natural Park. This atoll coral reef in the Sulu Seas is a marine sanctuary, with deep coral walls that are habitats for more than 1,000 species of fish, many of them endangered, and also sea turtles.

Shell-shocking

One of the starfish's few predators, the triton (*Charonia tritonis*), has become rare due to over-collection. This snail is unlucky enough to have a beautiful shell, which is used traditionally for decoration and even as a novelty musical instrument. Its drop in numbers may have triggered starfish swarms.

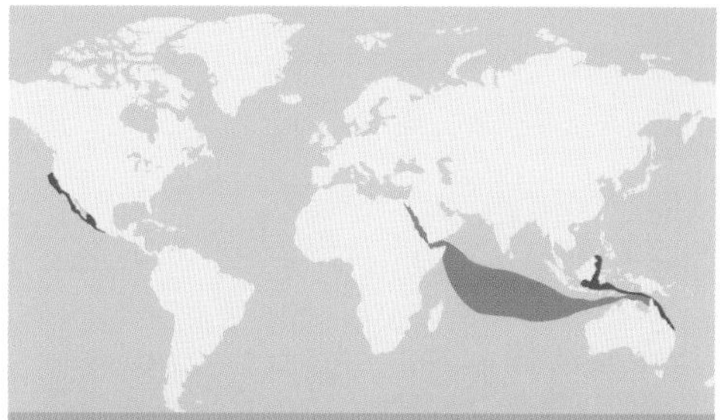

Fact file

- Adults reach up to 1m across and have as many as 20 arms, each with poisonous spines that can inflict very painful wounds.
- Tagged starfish travel more than 250m each week.
- The Japanese have used starfish bodies as fertiliser during mass infestations.

There is some evidence that crown-of-thorns attacks have been happening for more than 7,000 years, which leads scientists to wonder if the reefs are able to recover long term from an invasion. However, coral reefs today are already threatened by climate change, overfishing and other factors, and could do without the additional stress of an invasion by crown-of-thorns starfish.

In an average year, a female starfish can produce 60 million eggs, but this can increase to a staggering 250 million when conditions are right. One of the most likely causes of the crown-of-thorns population explosion is the run-off of nutrients from houses, farms and tourist resorts on to the reef. These nutrients provide masses of food for crown-of-thorns larvae, giving them a better chance of survival.

40. Comb jellyfish

Mnemiopsis leidyi

If you want an example of how a marine animal alien species can have a catastrophic impact on its adopted habitat, look no further than the comb jellyfish. This globular invader crossed the Atlantic from the Americas to the Black Sea in ship ballast water back in 1982. Flushed into its new home, the comb jellyfish soon settled in and began causing major problems. It's a gluttonous predator that can eat up to ten times its own weight every day. It preys on zooplankton and fish eggs, which meant it brought about the total breakdown of the Black Sea natural food chain from the bottom upwards.

Comb over

By the middle of the 1990s, comb jellyfish made up 90% of all life in the Black Sea, and had spread to the adjacent Azov Sea and the Sea of Marmaris. The biomass of comb jellyfish in the Black Sea alone was more than a billion tonnes, greater than the whole world's total annual fish catch. The fishing industries of the Black and Azov Seas have been virtually wiped out, with the loss of thousands of much-needed jobs. Fish numbers have plummeted, and dolphin and seal populations were devastated as their food supply disappeared. Now the comb jellyfish has moved into the Caspian Sea, and via the hulls of oil tankers is working its way around the Baltic.

The comb jellyfish traditionally lives in river estuaries, where salt levels vary according to the tide. That makes it very adaptable. It can also cope with big changes in temperature, which makes it able to colonise other parts of the world with ease. Another of its talents is the ability to reproduce at an alarmingly fast pace: when the temperature is balmy and food abundant, one comb jellyfish can produce as many as 3,000 eggs in a day. What's more, the comb jellyfish is a hermaphrodite, possessing both male and female sex organs. So you only need a singleton arriving in a new habitat to cause massive destruction.

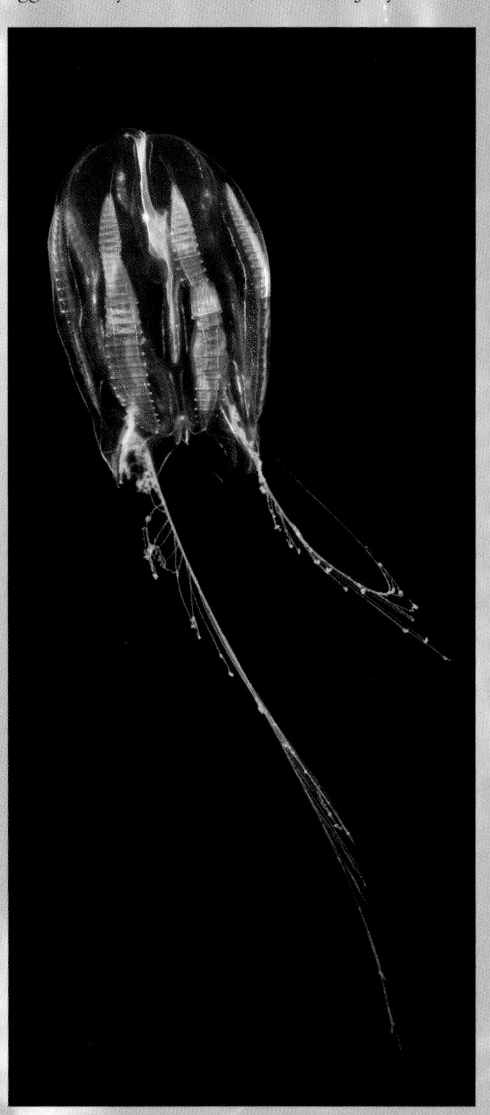

▲ *Shipping has helped comb jellyfish spread throughout the Black Sea, with devastating consequences for the fishing industry.*

Fact file

- **The comb jellyfish is not a true jellyfish. It belongs to the taxonomic phylum Ctenophora whereas 'proper' jellyfish belong to the phylum Cnidaria.**
- **It is exceptionally good at healing its wounds, able to repair damage to up to half its body.**
- **It is also known as 'sea walnut'.**

41. Caribbean barnacle

Chthamalus proteus

Barnacles are the world's greatest hitchhikers, always looking for a free ride. But the Caribbean (aka Atlantic) barnacle is bringing with it a whole lot of trouble. Larvae of this miniscule crustacean, smaller than a thumbnail, were taken on ship hulls and in ballast water to the Pacific, as recently as the 1970s. The free-swimming larvae rapidly attached themselves to any underwater structures they found, forming large colonies of adults that breed prolifically. Today, this tenacious barnacle is the most common organism in waters around Hawaii.

Darwin and the barnacles

Charles Darwin's reputation as a first-rate scientist was built on barnacles. His fascination with these tiny animals began during his voyage on the *Beagle*, with the discovery of a barnacle species that he nicknamed Mr Anthrobalanus (later classified as *Cryptophialus*). More than a decade later, Darwin was to spend eight years producing four volumes of drawings and classifications – this collection of material remains the most authoritative work about barnacles ever published. The time he spent identifying the many different varieties of barnacle helped him realise that they were descended from a common ancestor, an idea that was central to his theory of evolution.

Rocking out

The Caribbean barnacle is highly adaptable, spreading along coasts, estuaries and marinas and able to survive in temperatures from a chilly 16°C to a cosy 38°C. It fouls the local environment and takes over the territory of native rock-clinging species such as the Hawaiian limpet (*Siphonaria normalis*) and Hawaiian barnacle (*Nesochthamalus intertextus*).

Barnacles and other waterborne invasive species love the rubbish we throw into the sea just as much as they love ships. Studies have shown that alien species in the tropics breed three times as quickly in areas that have high concentrations of plastic waste.

Fact file

- **The Caribbean barnacle has 'arms' known as cirri, which it uses to scoop food from the water.**
- **Although it looks a bit like a mollusc, the Caribbean barnacle and its relatives are actually crustaceans, meaning they are more closely related to crabs and lobsters.**
- **Barnacle species such as *Fistulobalanus pallidus* (named by Charles Darwin) have been discovered attached to the leg rings of migratory seabirds.**
- **One species of barnacle, the one Charles Darwin nicknamed Mr Anthrobalanus, has the biggest penis of any creature in relation to its size.**

42. American signal crayfish

Pacifastacus leniusculus

It's oversexed, over-sized and over here. That's the complaint from wildlife groups and scientists right across Europe about the invasion of the American signal crayfish. This large-clawed, lobster-like crustacean was introduced into Scandinavia in the 1960s when the native crayfish (*Astacus astacus*) had been struck down by a crayfish plague (*Aphanomyces astaci*), and people were missing their crayfish suppers.

Signal failure

Sadly, nobody then understood that the American signal crayfish was not only a carrier of the plague, but was immune to its effects. The result was another dive in the fortunes of the luckless local crayfish. By the time scientists realised the magnitude of the disaster, the American signal crayfish had clawed its way into 25 European countries, including Britain and the Isle of Man.

The robust American not only passed on the plague to Britain's native white-clawed crayfish (*Austropotamobius pallipes*) but rapidly became a voracious predator of endemic species in rivers, streams and canals. Britain rushed through laws in the mid-1990s forbidding the keeping of exotic crayfish, but no legislation could halt the spread of the signals.

This cranky crustacean is a highly aggressive predator. It competes against trout and salmon, and devours smaller native British fish species such as the bullhead and stone loach, as well as frogs and invertebrates. With few natural predators this side of the Atlantic, it sometimes preys on its own species (nice). And it likes a bit of veg with all that protein, so causes a lot of damage to aquatic plants.

Besides being a greedy eater, the American signal crayfish's burrows cause significant damage to stream banks and canalsides. This raises the risk of flooding, and ousts highly threatened riverside animals such as Britain's water vole (*Arvicola amphibius*), the animal that inspired the much-loved character 'Ratty' in classic children's book *Wind in the Willows*. Little wonder the British government now encourages all anglers to kill these alien invaders on sight – though happily one quality that the American shares with its European cousins is its delicious taste, and thanks to recent TV cookery shows some Brits now use their alien crayfish for a cheap and eco-friendly dinner.

Fact file

- The American signal crayfish can grow up to 18cm.
- The female lays up to 400 eggs in the autumn, then carries them around under her tail until the spring.
- Its lifespan is up to 20 years.

43. Sea lamprey

Petromyzon marinus

This vampire fish native to the North Atlantic is sucking the life out of species in America's Great Lakes. The sea lamprey uses its mouth like a suction cup. It attaches itself to a passing victim, usually a fish larger than itself, then injects an anticoagulant to prevent the blood of its prey from clotting. It's a toss-up whether its victim bleeds to death straight away or dies soon after from infection.

Sucked dry

The sea lamprey was first sighted in Lake Ontario in the 1830s, probably introduced there via the newly opened Erie Canal. By the 1930s, it had swum along the Welland Canal into Lake Erie and was colonising Lakes Huron, Michigan and Superior. Within just two decades, the commercial fishing industry in the Great Lakes had virtually collapsed as sea lampreys devastated fish populations, altering the ecological balance of these inland waterways. They did so by attacking fish at the top of the food chain, such as trout, whitefish and lake herrings. The fall in the population of these big eaters left an eco-vacuum that allowed other invasive species to take hold.

Fish are not the only animals at risk from the blood-sucking lamprey. As recently as summer 2010, a cross-Channel relay swimmer was attacked by a sea lamprey. It stayed firmly attached until the swimmer had completed his stretch and returned to the pilot boat.

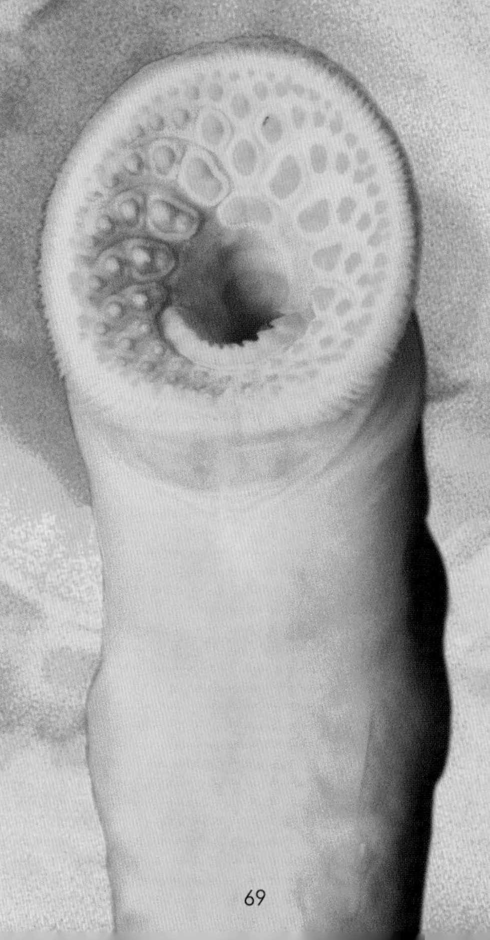

Shooting blanks

One of the most effective ways scientists can halt the invasion of sea lampreys is by trapping and sterilising the males. They're then released back into rivers to 'mate' with unsuspecting females who waste their eggs in these unsuccessful liaisons. More conventionally, fish and wildlife authorities are having some success controlling the sea lamprey by installing barriers along rivers and streams. These prevent the sea lampreys from spawning but allow passage for higher-jumping fish.

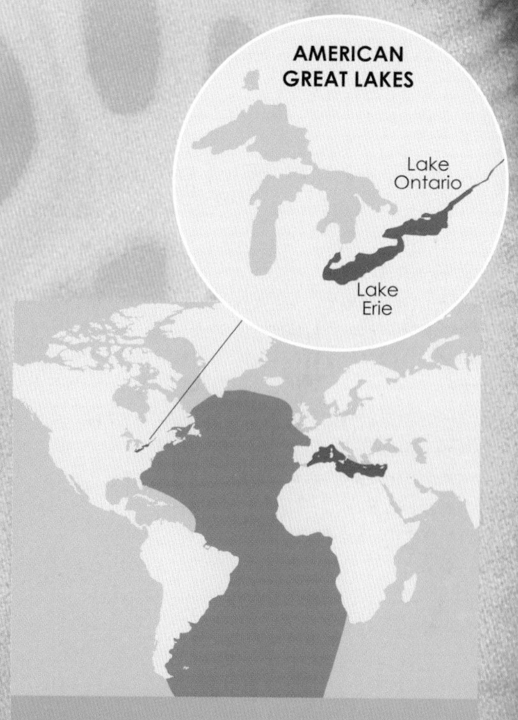

Fact file

- **King Henry I is said to have died after eating a 'surfeit of lampreys'.**
- **Fossil records of sea lampreys date back 340 million years.**
- **Lampreys are jawless fish, and also lack fins on the undersides of their bodies.**

44. Northern snakehead

Channa argus

In the year 2000, a man in Crofton, Maryland, ordered two live northern snakehead fish from a New York fishmonger, intending to make them into a nourishing broth for his poorly sister. However, by the time the fish arrived, his sister was well again, so he released the fish into a nearby pond. They bred – and the discovery that the pond was teeming with baby snakeheads sparked a national outrage and tighter laws against further releases.

Hungry horror

In 2004 it was discovered that this dangerous alien had also become established in the Potomac River – the fourth largest along the Atlantic coast – possibly from a separate release. It is now invading waterways as far south as Florida and the Mississippi Basin around Arkansas. Individuals have been caught as far away as California and – worryingly – the Great Lakes.

▼ *Vietnamese fisherman with a captured snakehead*

What those who let loose the snakeheads didn't realise was that this aggressive air-breathing monster can survive out of water for up to four days. It can wriggle on its pectoral fins to the nearest river but only after having eaten almost everything in its way. The northern snakehead can sweep a pond clean of native species, gobbling fish up to a third of its own length like native bluegills, largemouth bass and pickerels. It is having a devastating effect on wildlife in the Mississippi Basin, and snakeheads caught in the Potomac are found to have preyed on 15 different varieties of fish as well as crayfish, small birds, mammals and amphibians. A favourite with the more twisted end of the aquatic pet trade, a snakehead can devour six to eight goldfish in just a single day.

The northern snakehead does particularly well in North America because it has no natural enemies. It's highly adaptable, able to survive under ice in freezing conditions but also cope with water temperatures up to 30°C in the Deep South. Nor does it need as much oxygen as many American species, giving it a competitive edge over the native fish.

This alarming-looking beast has a snake-like head and a big, tooth-crammed mouth.

So frightening is this monster of the deep that it has inspired two sci-fi/horror films, *Snakehead Terror* and *Frankenfish*, both based on the release at the original Crofton pond.

Fact file

- Snakeheads often mate for life.
- This species is a delicacy in southeast Asia and India, often kept live in restaurant tanks. In America, supermarkets are prosecuted for selling snakehead.
- The longest snakehead recorded measured 1.8m, the size of an average human.

45. Lionfish

Pterois volitans

The story rolled out in every divers' bar from Bermuda to the Bahamas goes that the lionfish invasion in the Caribbean began when six fish escaped from a broken tank during Hurricane Andrew. They slipped into Florida's Biscayne Bay, and the rest is history.

Think tank

A more likely cause of the outbreak are releases by amateur aquarists, dismayed that their lionfish gobbled up everything else in the aquarium.

Lion hunts

Conservation authorities based in the Caymans enlist specially trained divers to catch lionfish. A lionfish hotline has also been set up in the Florida Keys National Marine Sanctuary so fishermen and divers can report sightings. More than 100 Keys dive operators have been trained and recruited to catch these toxic predators within the park. There have even been 'Lionfish Derbies' across the Caribbean and Florida with generous cash prizes for successful lion hunters.

Today, this venomous Pacific invader has spread right along the US southeast Atlantic coast and down through the Caribbean to northern South America. There have even been sightings as far north as New York, as this wannabe king of the sea expands its empire.

This striking fish gets its name from its impressive 'mane' of venom-tipped spines. And, like a lion on the African savanna, it tops the food chain in its new home, with neither adults nor their numerous eggs taken by other predators.

The invasion of the lionfish has been catastrophic for the health of the coral reefs in the Caribbean. It preys on 56 species, favouring juvenile and small fish and crustaceans, and already threatens several coral species with extinction. Studies indicate that the arrival of just one lionfish on a reef leads to a 79% drop in the number of young fish of other species within just five weeks.

Lionfish are efficient killers, and are extremely adaptable. They can take over artificial reefs, such as the 42m-deep Vandenberg in Florida, but are equally at home in the shallows or even mangrove swamps. They are also less susceptible to parasites and diseases than Atlantic and Caribbean fish.

Now America's National Oceanic and Atmospheric Administration is campaigning to convince the American public to eat lionfish, pointing out that they are extremely tasty and safe to eat, provided you remove the spines. Marine conservation charity REEF and website www.lionfishhunter.com have even published lionfish recipes.

Fact file

- **Despite concerns, the fish pet trade still imports lionfish – as many as 8,000 each year in Florida's Tampa Bay area alone.**
- **This fish can live more than 15 years in the wild.**
- **Though rarely fatal to humans, lionfish stings cause great pain, nausea and breathing problems.**

CORAL REEFS UNDER THREAT

A coral reef is like a lush rainforest on the edge of a much bigger desert. These reefs cover only about 0.2% of our oceans, yet are home to a quarter of all marine species. They're cradles for new plants and animals, sheltering more species than even the freshwater shallows, and they provide a livelihood to 200 million people in tropical countries. Every year, the coral reefs of the world contribute about US$30 billion from fishing and tourism. And they protect homes from the ravages of hurricanes and tsunamis. Low-lying islands such as those in the Maldives would be swamped if it wasn't for the protection of coral reefs.

▲ *Clown anemonefish play a vital ecological role on a healthy coral reef.*

On the way out?

Sadly, this will not be the case in decades to come if we don't find ways to protect these rich underwater ecosystems. At least a fifth of our coral reefs have already been lost, probably for good, and a further third of the corals that build our reefs will become extinct in the next century if we don't stop the rot. Climate change is the biggest problem, but alien invaders, run-off from coastal developments and farms, over-fishing, dynamiting and tourism are also putting immense pressure on these delicate ecosystems.

Climate change

Global warming caused by the build-up of carbon dioxide in our atmosphere has had a triple whammy impact upon our coral reefs. First of all, warmer sea temperatures destroy the algae that protect the coral, which then bleaches white and dies. The Great Barrier Reef in Australia is the world's largest reef ecosystem, with 2,900 individual reefs stretching collectively more than 3,000km. It's suffered extensive damage three times during the-called El Niño years, when the surface temperature of the southern hemisphere oceans rises.

The reefs around the Seychelles lost 90% of their coral cover during an El Niño weather pattern in the late 1990s. Happily, these reefs are beginning to recover, mainly because they're tough granite corals, able to support regrowth. But rising CO_2 levels are also increasing the acidity level in our seas. This slows down the growth of the coral, delaying its recovery. Bathwater temperatures are also great breeding conditions for the diseases that have killed off 80% of the Caribbean reefs in just two decades. The worst of these killers is black band disease, which has become the plague of the underwater world.

Rainforest and run-off

Felling rainforests for agriculture and tourism has a huge knock-on impact upon the world's coral reefs. Ancient trees hold on to the soil. When they're chopped down, the soil flows downriver to the sea, muddying the water and stopping sunlight reaching the reef. No sun means no coral growth. Such run-off from forest clearance is one of the main reasons why 60% of the coral around Hawaii's main islands has been lost.

Construction and farming also means more sewage and agricultural fertilisers going straight into the ocean. These raise the level of nutrients around the reef, feeding a new and dangerous generation of coral predators. Chief among them is the crown-of-thorns starfish (see page 65) which is threatening coral reefs as far apart as Mauritius, Australia and the Red Sea. Another relatively new threat to reefs in the Indian and Pacific oceans is the drupella snail, which feeds on living coral and leaves behind telltale white scars as evidence of its crimes.

Overfishing

Coral reefs are being overfished and large, valuable species such as groupers are disappearing fast. This disrupts the whole food chain. One of the most destructive forces is a new form of industrial fishing called bottom trawling, that began in the 1980s. Large rubber rollers attached to nets scour the rough surfaces of the reef. These cause huge damage. The Worldwide Fund for Nature reports reef scars up to 4km long on reefs in the northeastern Atlantic, and says this form of fishing has turned 90% of coral surfaces off southern Australia into bare rock. Fishing with cyanide for the aquarium trade is also causing widespread damage to coral reefs. For each live fish caught using cyanide, a square metre of coral reef is destroyed.

How you can help

Tourism is another threat to the health of the world's reefs, particularly walking on the reef and clumsy snorkelling and diving. Avoid doing harm by not touching or kicking the coral or stirring up sediment – and never drop anchor on a reef. Ideally, choose a dive or snorkel company that has a strong conservation policy and keeps a careful watch on its divers. Better still, combine a holiday with marine conservation work by volunteering with groups like Coral Cay Conservation (*www.coralcay.org*). Don't stay in coastal resorts that spill sewage and chemicals into the sea. Check hotel websites – if they don't state that they're actively looking after the reef, then they're probably causing damage. The souvenir industry is another source of reef damage. Never buy coral jewellery – the removal of red coral simply kills the reef.

▼ *Crown-of-thorns starfish feeding on coral in Komodo National Park, Indonesia*

46. Rainbow trout

Oncorhynchus mykiss

The release of farm-reared rainbow trout into North American waterways has had a dizzying effect on other species. Rainbow trout have been blamed for the spread of many ailments to wild fish, one of the worst being the parasitic 'whirling disease'. This causes curvature of the spine, the infected fish literally swimming around in circles as if chasing their tails.

Power and passion

The rainbow trout is the most commonly released fish in the world, popular as sport fish and also for the table. It is now present in at least 45 countries and every continent except Antarctica, and has brought about the decline and even extinction of many species. Apart from spreading disease to its wild salmonid relatives, this pretty fish is highly aggressive and competes for food and territory. A big bruiser, it grows up to 1.2m and weighs in at 25kg. It's also a complete carnivore, eating no greens at all.

Among the species at greatest threat from rainbow competition are the Colorado River system's endangered humpback chub (*Gila cypha*) and Australasia's small koaro fish (*Galaxias brevipennis*).

If the rainbow isn't fighting, it's getting smoochy, hybridising with other species and diluting their gene pools. Outside its native western North America, rainbow trout have almost totally seen off the Alvord cutthroat trout (*Oncorhynchus clarki*) by hybridisation. Other fish species on the unlucky-in-love list are the endangered Gila (*O. gilae*) and Arizona trouts (*O. apache*).

Threat to amphibians

Frogs and other amphibians generally come off second best when they the meet introduced rainbow trout, especially in naturally fish-free high-altitude streams and lakes. In the 19th and 20th centuries, angling clubs and even government fisheries stocked trout in these waterways, sometimes using aircraft to release hundreds of thousands of tiddlers at a time.

Sierra Nevada's yellow-legged frog (*Rana sierrae*) didn't stand a chance, as it had no evolutionary defences against fish. While its lowland cousins are armed with toxins and other anti-predator devices, the yellow-legged frog simply disappeared wherever trout were introduced. Today, it has vanished from 80% of its territory. The spotted tree frog (*Litoria spenceri*) of southeastern Australia has only ever been found in 21 mountain streams. Scientists report it has now disappeared from four of those streams, and blame introduced rainbow trout for feeding on the tadpoles.

Fact file

- **Not all rainbow trout swim out to sea. Those that do are called 'steelheads'.**
- **Steelheads return to spawn in rivers after two or three years in the ocean.**
- **A rainbow trout can leap up to four times its body length.**

47. Common carp

Cyprinus carpio

When Jesus talked about loaves and fishes, he could almost have been referring to the carp's miraculous ability to multiply. However, with a female laying a million eggs in a season, the carp doesn't really need divine intervention. The ancient Romans began shipping the carp across the known world in the year 1 AD – archaeological excavations have discovered Roman carp ponds along the shores of the Danube. Later, monks began farming carp and took the fish to western Europe to raise in stew ponds. Meanwhile, the Chinese have been cultivating it for 3,000 years – even today about three million tonnes of carp is sold in Chinese fishmongers each year.

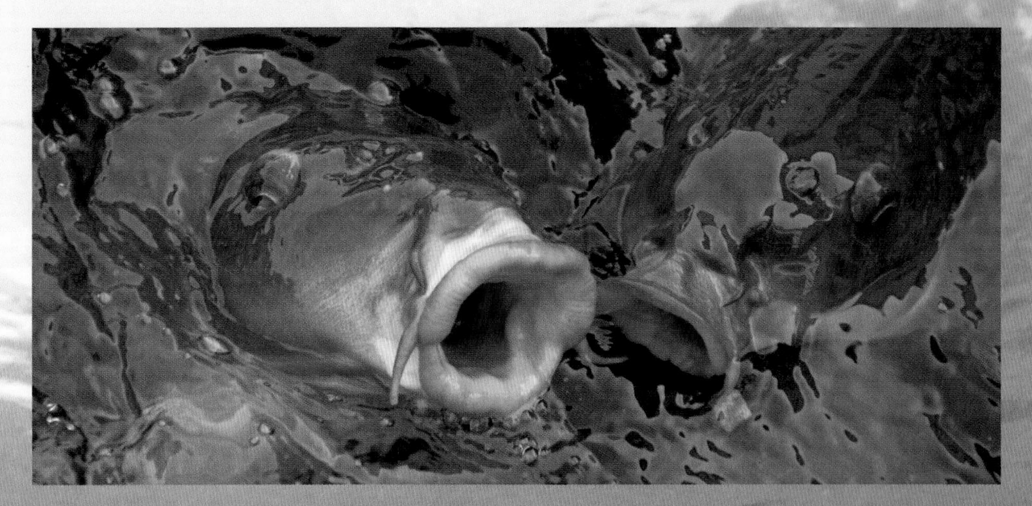

Carp arks

Through the 19th and 20th centuries, the carp was shipped as far afield as Australia, New Zealand and North and South America as a sport fish and for ornamental gardens. Today, it remains the world's most transported fish and has invaded every continent except Antarctica. It has also become a dangerous aquatic alien.

This portly fish is a bottom-feeder, churning up gunk on the riverbed that other fish would sensibly leave alone. This causes massive environmental damage, altering the chemical balance in the water and making it difficult for native species to survive. The carp sucks up all the debris then blows out anything not required. It is particularly problematic in shallow wetlands that are nurseries to other species, and also competes with native species for food such as crustaceans and plankton.

In areas such as the Murray River system in Australia, the carp is directly blamed for the huge drop in numbers of indigenous fish and waterfowl. The Australians are so worried by this that they permit carp to be caught year-round, with no restriction on numbers.

This invader is not going away without a fight. It's incredibly hardy, able to live in quite salty estuaries, in slow-flowing rivers or rapids, lakes and wetlands. It can cope in stagnant water with reduced oxygen levels, and even survives in polluted lakes and rivers. It's no tiddler either: a mature carp can weigh up to 40kg. Worst of all, it's spreading fast – tagged wild carp in Australia were recorded swimming distances of almost 900km at nearly a kilometre an hour.

Fact file

- Ornamental koi carp soon revert to their natural dull colouring when released into the wild.
- A carp lives up to 20 years in the wild – 50 in garden ponds.
- The world's oldest carp is believed to have been a koi named 'Hanako', who died in the 1970s aged 226 years.

48. Walking catfish

Clarias batrachus

▲ *A great blue heron does its bit in the battle against the walking catfish.*

If there's one thing worse than an aquatic invasive predator, it's an aquatic invasive predator that can walk. The walking catfish began life in the swamps and rice paddies of the Mekong and Chao Phraya river basins, where it has many natural predators. Unfortunately, it was introduced to many parts of the world as a farmed fish and become a huge threat to native species.

Walk of shame

This whiskery-faced, mud-coloured beast has a remarkable ability to thrive in conditions where other fish could simply not survive. Not only can it move across land to new habitats by shimmying along on its pectoral fins, it can also live out of water for days and can go months without food. The walking catfish is air-breathing and can tolerate stagnant waters where oxygen levels are low, such as muddy ponds and drainage ditches. The species is at its most lethal during droughts, when large numbers will walk to isolated ponds and eat everything in sight.

Walking catfish were accidentally released in Florida in the early 1960s and soon spread through the huge network of connected canals across the state. Within a decade, their population had soared, to the extent that more than 1,300kg of catfish was recorded in less than half a hectare in one area of southern Florida.

The walking catfish even invades Florida fish farms, where it causes thousands of dollars worth of damage to stock. In the Everglades it is a greedy consumer of native species, taking crustaceans and insects in good years, and muscling into temporary wetland pools in dry years to gobble up tadpoles and out-compete even larger species of native fish. Finally, it is surely the only fish that can be a hazard to road traffic: highways in south Florida can become slippery with catfish on wet nights as huge populations walk between ponds.

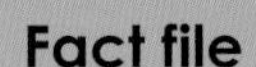

Fact file

- **The walking catfish's Thai name Pla duk dam translates in English to 'dull-coloured wriggling-fish'.**
- **Fish farmers in Florida build fences to keep it out of their ponds.**
- **A walking catfish found in the Thames, London, in 2009 was thought to have been a released pet.**

49. Nile perch

Lates niloticus

The gluttonous Nile perch is a big bully-boy, weighing in at about 230kg. It was introduced to Lake Victoria in the 1950s to create a commercial and sport fishing industry, and with no natural predators went straight to the top of the food chain. Within five decades the Nile perch had seen off an estimated 200 species of fish in Africa's largest lake. The losers were mostly freshwater fish called cichlids, in particular the endemic haplochromine cichlids.

Hungry for more

Lake Victoria's whole ecology soon began to change. The massive decline in the population of algae-eating fish caused the algae to grow and oxygen levels to fall – bad news for all species dependent on the lake, such as the resident pied kingfishers (*Ceryle rudis*), which feed on cichlids.

With such a hefty weight to support, the Nile perch will eat just about anything in its path – zooplankton when it's a baby, then insects, crustaceans and molluscs. As it grows, it eats bigger and bigger prey. Now, the perch is running out of other species and even turning on its own kind. Yet it continues to breed terrifyingly well. Each female produces an average of nine million eggs in her lifetime, and while many individuals don't make it to maturity, a grown Nile perch can live for up to 16 years.

The Nile perch problem goes beyond the lake itself. The fish's flesh has a high oil content and needs to be smoked as part of the preserving process. This has increased demand for fuel in Uganda, Kenya and Tanzania, and led to widespread deforestation and erosion.

Breaking traditions

Local communities have not benefited much from the opening of fish factories around Lake Victoria. Many traditional village fishermen were put out of business by the new perch entrepreneurs. Most local people relied on traditional lake fish such as catfish as their main source of protein. These have disappeared, and traditional nets are too flimsy to catch the huge Nile perch. The export market has also driven the price of Nile perch beyond the reach of most locals, not that many would want to eat it anyway. The destruction of the lake and loss of the traditional catch has left a bad taste in everybody's mouth.

▲ A Nile perch swallows a baby Nile crocodile.

Fact file

- **The Nile perch was revered in parts of ancient Egypt and sometimes mummified.**
- **It's born male. Some become female when they reach sexual maturity at three years old.**
- **It's also known as the 'African snook'.**

THE EXPLORERS

When Lieutenant James Cook set sail for the South Seas on board the *Endeavour*, he had little idea that his expedition would radically change the way we understand science and nature. Secret orders opened during the 1769 expedition told Cook to sail south in search of 'Terra Australis Incognita', and bring back as much information as possible about the animals, plants and minerals of this unexplored land.

▲ Tui on New Zealand flax

Voyage of discovery

On board were scientist Joseph Banks and his assistant Daniel Solender, who had been a pupil of Carl Linnaeus, the greatest botanist of the 18th century and the inventor of the biological classification system that we use today. Banks, Solender and their team collected 3,000 plant species, more than a third of which were completely new to science.

The *Endeavour* scientists were among a remarkable group of explorers from Magellan to Charles Darwin who transported plants and animals backwards and forwards between the Old and New Worlds. Some of these species brought huge benefits, such as the potato, taken to Spain from South America by the Conquistadors and to England by privateer Sir Frances Drake. Joseph Banks discovered New Zealand flax (*Phormium tenax*) on the 1769 voyage and brought it to Europe, where it became important in the textile and rope industry. He also arranged the introduction of the Pacific breadfruit (*Artocarpus altilis*) to the British West Indies, where it remains a staple food in the Caribbean (see box).

However, certain other species discovered and moved to other countries by the great explorers caused enormous damage, and contributed to the plummeting of biodiversity in our modern world. One plant that has seriously outstayed its

Banks and the *Bounty*

The voyage of HMS *Bounty* was one of the world's first experiments in the large-scale shipping of a food crop from one part of the world to another. It was commissioned by Joseph Banks, who believed breadfruit would supply cheap but nutritious food for the slaves of the Caribbean. He had the *Bounty* hold lined with copper to stop marine worms attacking the saplings, and hired Kew gardener David Nelson, who'd sailed with Cook on his third and fatal expedition to the South Seas. In charge of the voyage was another seafarer from that luckless expedition, the young Lieutenant William Bligh. The infamous mutiny on the *Bounty* failed to curb the British government's enthusiasm for breadfruit. A second voyage led by the newly promoted Captain Bligh in 1791 succeeded in establishing breadfruit forever in the Caribbean.

▼ Breadfruit

Guanaco in Torres del Paine National Park, Chile

Magellan and the camel without humps

Our fascination with the discovery of new species owes much to the ship's log of adventurer Ferdinand Magellan. He lead the first expedition to circumnavigate the globe in 1519, and his discoveries brought about an explosion in the spice trade between the Far East and Europe. Although Magellan was among the 232 men and four ships lost during the voyage, the surviving ship *Victoria* limped into a Spanish port three years later with enough spice to turn a handsome profit for its owners.

However, while spice excited the merchants, it was the reports of strange animals in exotic surroundings that captured the imagination of the Spanish people. Such was their thirst for knowledge that future explorations would be as much about discovering the natural world as about conquest and trade.

One of the species in Magellan's log that caught their attention was the 'camel without a hump', an animal that is probably the guanaco (*Lama guanicoe*), a member of the camelid family that lives around Tierra del Fuego. Then there were the tasty black geese that the voyage cooks complained had to be skinned rather than plucked. Today, we call them penguins, and it's rather fitting that the Magellanic penguin (*Spheniscus magellanicus*) takes its name from the legendary explorer.

Charles Darwin

welcome is eucalyptus, brought home from the newly charted Australia by Banks. Ships such as *Endeavour* and Darwin's *Beagle* also undoubtedly introduced rats, cats, pigs and goats to isolated islands.

Many of the species recorded by those explorers are now extinct. More than half of the 140 endemic birds collected on Hawaii during Cook's *Discovery* expedition no longer exist. Thankfully, the meticulous records and drawings made by intrepid scientists are giving us an insight into the genetic evolution of disappeared species and their surviving cousins. Today, there are 738 drawings and paintings by *Endeavour* artist Stanley Parkinson in Britain's Museum of Natural History. That sort of detailed information is as relevant in the 21st century as it was 240 years ago, and is helping scientists protect threatened plants and animals today.

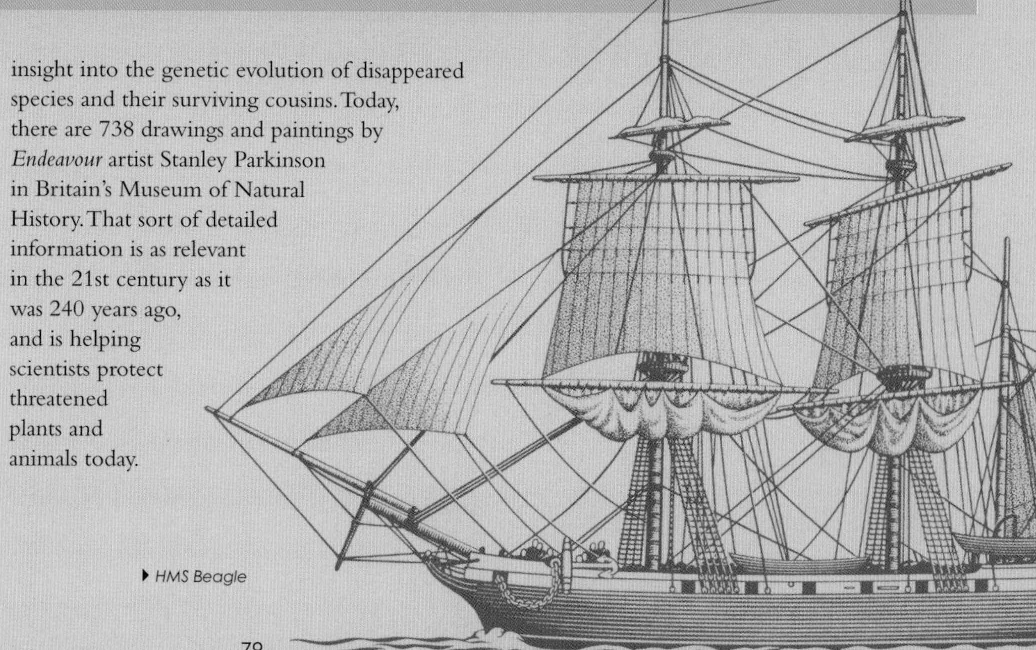
HMS Beagle

4 MARCH OF THE CREEPY-CRAWLIES

Insects and other invertebrates account for nine out of ten species on the planet. Often tinier than a pinhead, invasive bugs are able to topple giant forests and sabotage a nation's economy.

Imported red fire ants gather on the surface of their nest.

Most life forms on Earth are hardly bigger than a thumbnail. Some 90% of all animal species are invertebrates, meaning they don't have a backbone. These include spiders, worms, crabs, all those squashy molluscs and, above all, insects. They may be spineless but they're the backbone of every ecosystem – pollinating plants, aerating and fertilising the soil, and providing food for bigger animals. Many friendly insects also keep nastier pests under control, thus protecting our food crops. Entomologists believe there are probably more than eight million different species of insect on the planet.

▲ *Ghost forests in Great Smoky Mountains National Park, devastated by the balsam woolly adelgid*

Bug invaders

You don't have to be big to have an enormous impact on the environment: the little Asian longhorn beetle (see page 86) can topple ancient forests. Alien bugs can tip the balance of an ecosystem, with grave global and economic consequences.

Most of the world's invasive insects arrive as larvae in the soil or attached to the foliage of exotic plants. One such disastrous import was the balsam woolly adelgid (*Adelges piceae*). This small wingless insect was brought into North America at the turn of the 20th century, and has devastated native balsam fir and Fraser fir forests, which had not developed natural defences against the new invader. The worst damage was in the Great Smoky Mountains National Park, where 95% of the Fraser firs were killed off, leaving what are now called the ghost forests.

Exotic creepy-crawlies now infest fragile and biodiverse habitats all over the world. One in four insect species in the Galápagos is introduced, with nearly 500 alien bug species, from wasps and fire ants to parasitic flies. In New Zealand there are more than 2,000 species of invasive invertebrates, including the aggressive common wasp (*Vespula vulgaris*) and 28 species of ant.

◀ *The common wasp, native to Europe, competes aggressively with indigenous species in New Zealand.*

Rogue agents

Bugs have also been deliberately introduced to new countries and ecosystems as biological controls, often with disastrous results. Biological control means introducing natural enemies to control insect or plant pests. It can work extremely well. However, sometimes the new aggressor ignores the pest it was brought into control and becomes an even bigger problem. A classic case is the introduction of Asia's harlequin ladybird (see page 87) into Europe and North America, where it has gone from being an aphid-eating farmers' friend to a serious threat to native beetle species.

Another disastrous introduction was that of the gypsy moth (*Lymantria dispar*) into North America, by French scientist Etienne Leopold Trouvelot. He introduced the moth from France in the 1860s as part of an experiment to create a silkworm industry in Massachusetts. Some of the larvae escaped, and by the 1880s they had become a major problem in the state. Today, these moths cause millions of dollars' worth of damage to America's native hardwood forests every year.

Bugs as victims

The world's bugs are under threat, from habitat loss, climate change and pesticides as well as pressure from invasive species. Bee populations are crashing – plummeting almost by half in France alone between the late 1980s and the turn of the century. This is a serious problem for all life forms, including us, because bees are the main pollinators of so many important plants. As Albert Einstein warned: 'If the bee became extinct, man would only survive a few years beyond it.' Butterflies are also at considerable risk. Globally, about one in ten species is near extinction and in Europe nearly a third of the region's 435 butterflies are in decline.

▲ A giant weta eating a carrot, Little Barrier Island, New Zealand

Wetland drainage is also having a serious impact upon insects such as dragonflies. In countries seriously affected by global warming, more and more water is taken from streams and rivers for agricultural irrigation – or even for swimming pools in tourist resorts. This is causing the rapid decline of many dragonflies, such as the Greek red damsel (*Pyrrhosoma latiloba*), which scientists fear may be extinct by the middle of this century.

Helping creepy-crawlies

▸ Gypsy moth caterpillar

Much can be done to help invertebrates at risk. Cutting down on the use of broad-based pesticides is a big step forward in protecting the world's invertebrates. Countries are also increasing biosecurity to stop invasive species being spread between habitats. Invertebrates respond well to breeding and release programmes because they don't need as much space as larger, backboned animals to thrive. They can also survive if we protect tiny fragments of their original habitat.

In New Zealand, scientists are reporting success in their efforts to save the giant weta, one of the largest bugs on Earth and heavier than a small bird. This enormous creepy-crawly evolved alongside native predators such as birds and reptiles, but had no time to prepare its defences against introduced cats, rats and hedgehogs. Now it is fiercely protected and recovering well as predators are cleared from its cave and forest homes.

50. Asian tiger mosquito

Aedes albopictus

It has become one of the fastest-spreading invasive aliens on Earth in the last 20 years, causing misery wherever it settles. The Asian tiger mosquito is an old enemy of man in its native southeast Asian, Pacific and Indian Ocean territories, a biter by day as well as the dawn and dusk preferred by its many relatives. A single mozzie can bite up to 48 times in an hour, carrying disease between people and other species. This species has been blamed for an epidemic of the debilitating illness chikungunya on the Indian Ocean island of Reunion, and for the spread of dengue and west Nile fevers.

Bloodsucker

The Asian tiger mosquito can only fly about a kilometre by itself so it relies on our help to cross wide oceans. It is highly adaptable and once established will compete aggressively with native species for food (that's us, and other warm-blooded creatures) and breeding grounds.

It breeds naturally in the pools of water that collect in tree branches, but will adapt to the smallest manmade pond, such as an old tin can or dog's drinking bowl. Its favourite nursery is a discarded tyre – and when tyres are transported on trucks or ships, the mosquito spreads. In 1985, it arrived in Houston, Texas, concealed inside a shipment of tyres.

Now it is a common pest right across the southern US and up the east coast as far north as New England. The striped invader also hitches a ride with the so-called 'lucky bamboo' – lucky for the mosquito, perhaps, but bad news for people. Lucky bamboo, aka *Dracaena* spp, is shipped from China mainly to Asian buyers in California, which means the insect is likely to become established on the west coast of the United States.

Mozzies and mankind

Mosquitoes are our deadliest animal enemies, and have killed more people than all our wars combined. Malaria alone kills up to three million (with 200 million people infected) every year. It's spread by the *Anopheles* mosquito family. Millions more people are infected with other mosquito-born diseases such as yellow fever, dengue or encephalitis, and these are all spread by the Asian tiger mosquito.

The Asian tiger mosquito easily adapts to subtropical and temperate conditions worldwide. As the planet warms up, scientists fear that warmer weather will lead to a greater spread of mosquito-borne diseases.

Fact file

- **Only the females bite. They need blood to develop their eggs. Otherwise they feed on nectar and other sweet plants.**
- **The larvae of all mosquitoes live in water.**
- **Asian tiger mosquitoes prefer to feed from mammals but will bite birds if we're not around.**
- **Mosquito means 'little fly' in Spanish.**

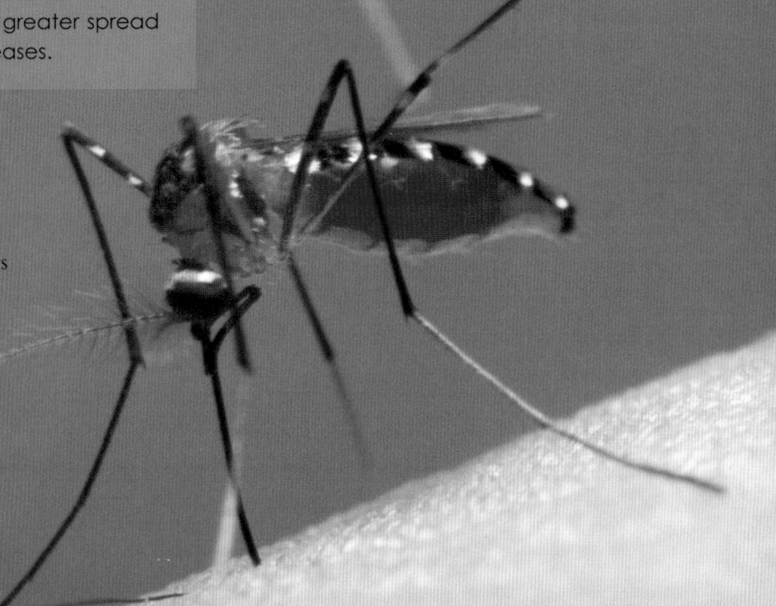

51. Formosan subterranean termite

Coptotermes formosanus

It may be tiny, but the Formosan subterranean termite, from southern China, does up to US$2 billion worth of damage every year in the US alone. This aggressive insect hitched a ride on cargo ships bringing timber into New Orleans after World War II. It took a liking to its new habitat, cheerfully chomping its way through the timber-framed houses of the historic French Quarter. It was soon out-performing the local termites, devouring timber nine times faster and building colonies ten times bigger than the more laid-back natives.

Termite time bomb

North American termites make colonies underground, whereas the Formosan subterranean termite can live above or below the earth – and moist, warm wooden ships, balconies and even high rises are all considered des res for this advancing alien. As well as causing extensive damage to houses and buildings, it has also been hollowing out live North American trees such as American ash and pecan.

The Formosan invader can escape detection for years. The first you know about a colony might be when your foot goes through a rotten floorboard. Then there are the late spring swarms when the winged adults (called 'alates') flit about to find a mate and establish a new colony. The Formosan subterranean termite uses people to help spread its territory, hiding away in pot plants, mulch, recycled timber and vehicle tyres.

The biggest aid to its expansion is the transport of railway sleepers, no longer needed by the railroad companies and sold to unwitting landscape gardeners across America.

Now these termites are in 11 states and America is fighting back. The National Formosan Subterranean Programme has been set up, aiming to find ways to reduce the termite population. Already there have been some steps forward. Scientists have discovered a new strain of infectious fungus called *Metarhizium anisopliae*, along with toxins extracted from wild celery, knapweed and mugwort. Any or all of these might put a stop to the termite sometime in the future. In the meantime, don't lean too hard against the balcony rails.

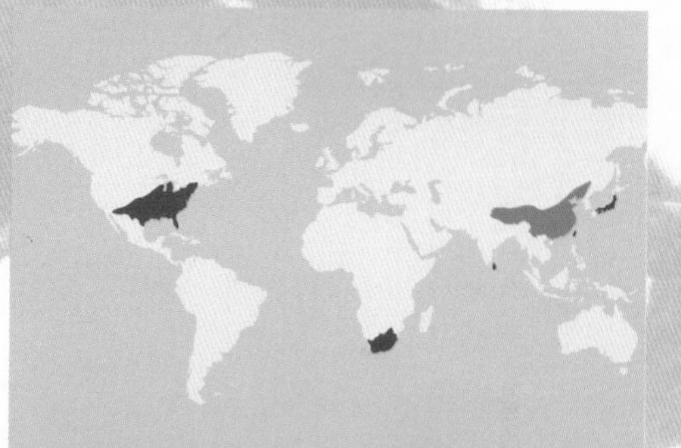

Fact file

- The colonies eat non-stop and consume about 400g of wood a day.
- In North America, termites do more damage than hurricanes and tornadoes.
- The total weight of termites is greater than the weight of all the people on earth.
- Formosan termite colonies under the ground can be almost 100m long.

52. Asian longhorn beetle

Anoplophora glabripennis

You don't need a chainsaw to fell a hardwood tree: the Asian longhorn beetle can topple an ancient forest as efficiently and ruthlessly as a team of loggers. This Chinese native arrived in the United States and Austria in wooden packing cases and pallets. The authorities were caught on the hop: Asian longhorn beetles had already been identified as a problem in their own part of the world.

Chainsaw-less massacre

In the Chinese province of Ningxia, some 50 million trees grown as part of an afforestation scheme in the 1980s had to be cut down because of severe damage by beetles. Now, this pretty little insect is a serious threat to America's hardwood forests, particularly in the eastern US, where a third of trees are at risk. The estimated replacement value of these trees comes to more than US$600 billion. The adult beetles bore into a tree, eating leaves and twigs. Some of the worst damage, though, is done by larvae wriggling beneath the bark and breaking down the tree's water and nutrient transport systems. Within five years of the beetles' arrival, the tree is probably dead.

At a time when the planet needs all the trees it can get, the Asian longhorn is a serious threat to our global environment. It also has a huge economic impact on timber-related industries, such as the production of maple syrup, and the tourism built around New England's spectacular autumn colours. At the moment, beetle attacks in Austria and the US have been contained within cities and surrounding parkland, where 'beetled' trees are cut down, chipped and burnt, and quarantine areas created around the infestation.

However, if the Asian longhorn beetle spreads into North America's native forests, it could bring about an ecological disaster of catastrophic proportions. So concerned are the American and Canadian authorities that an eye-watering US$800 million has already been spent on trying to contain this dangerous alien.

Fact file

- The longhorn's impressive banded antennae are as long as the whole body in females, and twice as long in males.
- It is known as the starry sky beetle in China, because it is black with white speckles.
- An Asian longhorn beetle can fly nearly 400m to colonise new trees.

53. Harlequin ladybird

Harmonia axyridis

This is the most invasive ladybird species on Earth and – in contradiction to the nursery rhyme – just won't 'fly away home'. The harlequin ladybird is threatening the survival of more than 1,000 native insect species in the British Isles, by eating their food and attacking their young. It took less than five years to spread across the whole of England after having been spotted in a pub garden in 2004. The highly invasive grey squirrel took a century to perform the same feat.

Unsuitable diet

The harlequin is the most widespread member of its family in North America, despite having only been successfully introduced there during the 1980s. An east Asian native, it was introduced to attack the aphids that were destroying crops. Unfortunately, the greedy harlequin hasn't confined itself to aphids, but happily dines on a wide range of other insects, and plants too. That gives it a competitive edge over more choosy, aphids-only ladybirds.

Everywhere it has been introduced, the gluttonous harlequin has quickly turned on native species that are vital to maintaining biodiversity. Its prey ranges from the larvae and eggs of butterflies and moths, to other species of ladybird. British scientists believe the 46 native species of ladybird, such as the two-spot (*Adalia bipunctata*), are now at serious risk.

The harlequin has a longer breeding season than other ladybirds, such as North America's native *Coleomegilla maculata*, which gives it another Darwinian leg-up in the battle for bug supremacy. The result has been a catastrophic spread of harlequins north and southwards, and a decline in America's endemic ladybirds and other native aphid-eaters.

Not content with unbalancing the American ladybird world, the harlequin is also a commercial pest. Although it has sorted out the aphid plague on pecan trees, it's invaded other crops that it was brought in to protect. Harlequins attack apple, pear and grape crops, spoiling wine when they are crushed up with grapes after harvest.

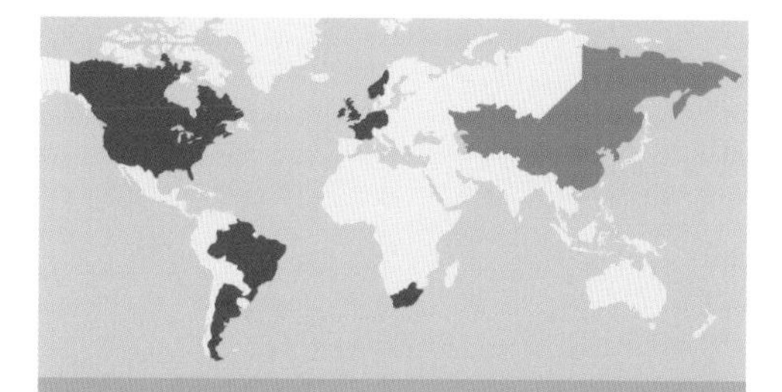

Fact file

- The name 'ladybird' dates back to medieval times, when these insects were called the 'Beetles of Our Lady'.
- In Britain, the public is being asked to report sightings on the website *www.harlequin-survey.org* so scientists are able to map the harlequin population.
- This species is very variable in appearance – some are black with red spots, others the reverse; and the number and size of spots vary too.

Harlequins also invade homes, swarming inside when the weather turns chilly and hibernating within the walls. But scientists are not giving up on the battle against the invading beetle. One of the latest proposals is to introduce a sexually transmitted disease into the Harlequin population that renders them infertile. This may sound extreme, but the behaviour of these ladybirds has been anything but ladylike.

54. Cannibal snail

Euglandina rosea

It is fast (for a snail) and deadly, and leaves a trail of death and destruction in its slimy wake. This is the cannibal snail, aka rosy wolfsnail, a Florida native that has been blamed for the extinction or rapid decline of other varieties of snails, wherever it has been introduced.

Gastropod gastronomy

The cannibal snail was deliberately imported to Hawaii in the South Pacific back in the 1950s, to deal with another invasive predator, the giant African land snail. Unfortunately, like all bullies, it was reluctant to take on a bigger rival, instead turning on the smaller, indigenous snails. One victim was Hawaii's O'ahu tree snail, extinct within just one year. On the Hawaiian island of Kauai, all 21 of the endemic *Carelia* snail species are believed to have vanished, and others are now on the Endangered list.

The lesson still not learnt, cannibal snails were taken across the Pacific to French Polynesia in the 1970s, with equally devastating results. The burly, 76mm-high cannibal snail soon developed a taste for the native tree snail *Partula faba*. In two decades, the native snail population on the island of Raiatea had dropped by 80%. Hopefully, fast intervention by Bristol Zoo has been able to save the species from extinction. The zoo is breeding the tree snails in captivity, caring for 88 precious little snails in a climate-controlled lab. Their offspring will one day be released into protected areas of the Pacific.

The slimeball cannibal snail is now colonising the tropical world. According to Columbia University, it now slithers far beyond Hawaii and French Polynesia to Samoa, Vanuatu, Papua New Guinea, Japan, Madagascar and the Seychelles, India, Sri Lanka and right across the Caribbean.

True cannibals

This snail will eat its own species as eagerly as it consumes other snails. Babies start devouring their weaker siblings as soon as they hatch. Unsurprisingly, adult cannibal snails tend to be solo animals. They put dinner on hold to mate and, being cross-fertilising hermaphrodites, both partners produce eggs.

Fact file

- The cannibal snail breeds five times faster than most other snails. An average adult will lay up to 40 eggs each year.
- It follows the slime trail of its prey, slithering up to three times faster than its victim.
- It will eat smaller snail whole, or launch a ram-raid attack on the shell of a larger mollusc.

Hawaii

Samoa

Seychelles

55. Glassy-winged sharpshooter

Homalodisca vitripennis

Tahitians rue the day that the glassy-winged sharpshooter was first spotted in an island garden back in 1999. Alerted by a sharp-eyed gardener, authorities were swift to kill the sharpshooter and burn the tree. They were too late. Although they acted immediately, they failed to stop the invader in its tracks. Just a few years on, France's Society Islands are overrun with the insect – thriving in far greater numbers than it does in either its native Mexico or even its other invasive territory, California.

Suck it up

It might have a great name, but this is one seriously troublesome creature. The glassy-winged sharpshooter is a large, 12mm-long leafhopper that causes enormous damage to fruit crops in California, and is shaping up to become an ecological nightmare in French Polynesia.

This handsome sap-sucking bug feeds on more than 70 different plant species and has a particular fondness for grapes, citrus fruits, almonds, stone fruit and oleanders. It feeds by inserting its viciously sharp sucking-tube of a mouth into the plant. As it sucks, it simultaneously squirts from its anus droplets of a nasty waste fluid known as leafhopper rain, which splatters anyone unlucky enough to be standing beneath the tree. Locals say that a picnic in Tahiti's leafy parks is no longer an option.

Mass feeding alters the water and nitrogen balance of the trees, making them susceptible to disease. In Tahiti, naturalists are concerned for the health of native plants. The glassy-winged sharpshooter has no predators in its adopted territory and can feed all year round rather than just seasonally in its native land. It has colonised the remaining areas of native forest in the islands, as high as 1,400m above sea level.

In California, glassy-winged sharpshooter nymphs carry a lethal plant bacteria called *Xylella fastidiosa*, which threatens the health of the state's US$5 billion annual grape industry. The bug is also blamed for passing on many other plant diseases, such as oleander leaf scorch.

The march of the glassy-winged sharpshooter continues. By 2004 it had arrived on Hawaii and a year later was spotted on Easter Island. This is not a bug that's going to be easily squashed.

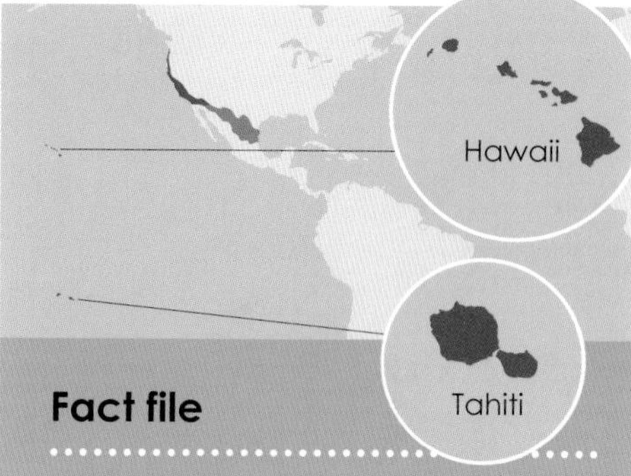

Fact file

- Despite biological security checks, glassy-winged sharpshooters have been found on aircraft flying from Tahiti to New Zealand, Australia and France.
- Control methods under investigation include a sharpshooter-specific virus, and the use of parasitic fairyflies – tiny wasps that attack sharpshooter eggs.
- Sharpshooters get their name from the force with which they expel their bodily waste.

56. Earthworm

Lumbricus terrestris

When the Pilgrim Fathers landed on the shores of North America, they brought with them a dangerous invader – the common earthworm. The worms stowed away in the ballast on board sailing ships. At the end of each voyage, the captain would order the ballast to be adjusted to suit his new cargo, and tons of soil – complete with worms – would be dumped on shore. In just 200 years the common earthworm had colonised the entire continent of North America, and today there are at least 15 species of earthworm wriggling their way across the northern states and Canada. Scientists estimate there may be up to 1.7 million worms per acre of land!

Woodland wastelands

The same friendly creatures that help till the soil of European gardens now seriously threaten the survival of America's hardwood forests. The European earthworms have voracious appetites, and rapidly chomp their way through the dead leaves and plants on the forest floor. They leave the forest floor bare, starving the young trees of nutrients and causing soil erosion. The run-off water from these forests also damages freshwater lakes and rivers, harming the fish. Who'd have thought a small, wiggly worm could be responsible for such havoc?

In agriculture, earthworms help farmers by creating tunnels that allow water to seep down, and they help fertilise the soil. The material they digest contains nutrients manufactured by plants during photosynthesis.

On a typical farm, the weight of the earthworms beneath the surface is greater than that of the farm animals on land. Each year earthworm castings (poo) cover each American acre with as much as 18 tonnes of rich soil. As Charles Darwin said, 'It may be doubted whether there are many other creatures that have played so important a part in the history of the world.' However, even in farming earthworms can do some damage – their castings create problems by increasing erosion along irrigation ditches.

How you can help

Even though they've spread fast across America worms still only move at about half a mile in a century. So don't help them spread further. Be careful about buying baitworms and potted plants, and don't dump soil.

Fact file

- **An earthworm eats at least half its body weight each day.**
- **It is ready to breed within six weeks of emerging from its cocoon, and may live for 50 years.**
- **Earthworms have been found two miles below the surface of the Earth. Down there temperatures can reach as high as 71°C.**

57. New Zealand flatworm

Pterois volitans

A faithful friend is missing from the gardens in Northern Ireland. It's the humble common earthworm, otherwise known as *Lumbricus terrestris*, and it's been gobbled up by a cannibal from Down Under. The New Zealand flatworm first appeared in a Northern Irish garden in the early 1960s and has gone on to devastate the earthworm population in the province. Now, this worm has chomped its way across Europe to mainland UK, Iceland and the Faroe Islands.

Sterile and starving

Earthworms aren't the only losers. The quality of soil is affected when the earthworm population declines, because less water can enter the ground through wormholes. Songbirds start to go hungry – earthworms being a favourite diet for young birds – as do moles, shrews and frogs.

The exotic plant trade is to blame for the introduction of New Zealand flatworms and it's not surprising that the biggest populations on the British mainland are around garden centres. Find a big shop that has been selling New Zealand flaxes or palms since the 1960s, and the chances are there'll be flatworms in the gardens nearby.

This worm extends up to a ribbon-like 17cm long when it is on the move, though you're more likely to encounter one in its coiled-up, resting state, perhaps sheltering under a stone. Adults are glossy purplish brown, but young specimens are pale pinkish, confusable with normal earthworms at first glance. Unlike the native earthworms, however, a flatworm is pointed at both ends and has no segments. It also has no natural predators, which makes it difficult to control.

If you find a flatworm in your garden, there are probably more lurking out there. You can round them up by placing an empty sack on the ground and weighing it down – check regularly for more wormy visitors. Look out for the egg sacs in summer: they are round and shiny, nearly 1cm long and dark red when fresh – though darken to black. Left alone, each sac will spawn up to ten more earthworm-munching baby flatworms.

Fact file

- If you see a flatworm in your garden, report it to the Scottish Crop Research Unit, who are studying the animal and its spread.
- It can survive without food for up to two years. It will just get flatter.
- A New Zealand flatworm moves at 17m per hour.

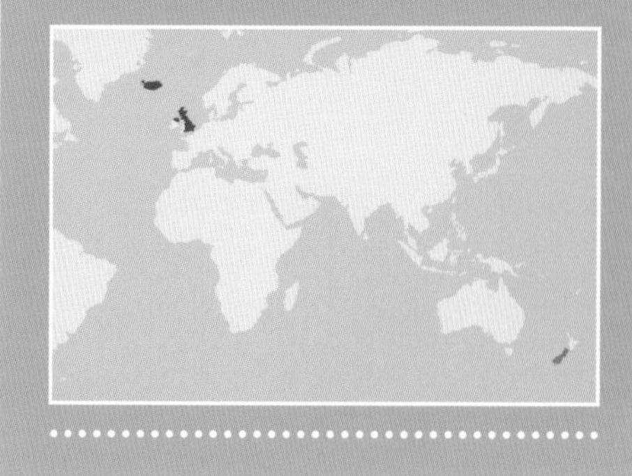

58. Desert locust

Schistocerca gregaria

Ever since Moses brought down the plague across Egypt, the desert locust has been one of the most feared enemies of mankind. Turning the sky black and extending several kilometres across, a swarm can put tens of millions of people at risk of starvation. In a swarm year, locusts can cover 30 million kilometres across 60 countries: that's a fifth of the planet. A tonne of locusts – a tiny percentage of a swarm – eats as much food as 2,500 people in a single day. A full swarm containing 50 billion locusts will devour 80 tonnes of crop in a day.

All you can eat

Ironically, it's the rain so badly needed by farmers in northern and western Africa that brings out the swarm. A female will only lay eggs after a heavy rainfall turns the countryside green, providing enough food for her newly hatched infants. This is an insect that definitely lives to eat, chomping its way through its own body weight from birth to death. Everything veggie is on the locust menu, from crops and fruit to leaves, grass, flowers and trees.

Early warning schemes

Like a weather report, the UN-run Food and Agriculture Organisation locust warning programme advises countries and farmers of outbreaks. It is backed by emergency response units who use pesticides to stamp out young colonies before they take to the skies. Once the swarm is moving, there is little anyone can do stop the invasion. Even today the FAO admits: 'There is little evidence that chemical controls – as opposed to winds, rain or lack of food – have wiped out plagues.'

Once in flight, a swarm is like a runaway train, travelling thousands of kilometres. Locusts devastate the area of Africa known as the Sahel, encompassing Mali, Niger, Senegal and Mauritania in northwestern Africa, across to the Red Sea. In bad swarm years, they travel up the Arabian peninsula into Iraq, Iran and onward through India, devouring all in their wake.

In 2004, a locust swarm consumed the food supply of more than 2.5 million households in the Sahel area. In Mauritania, one of the poorest countries on Earth, 80% of all food crops were destroyed. In 1954, desert locusts invaded the British Isles and in the autumn of 1987, plagues of locusts crossed the Atlantic, travelling 5,000km in just six days to descend upon the Caribbean and Latin America.

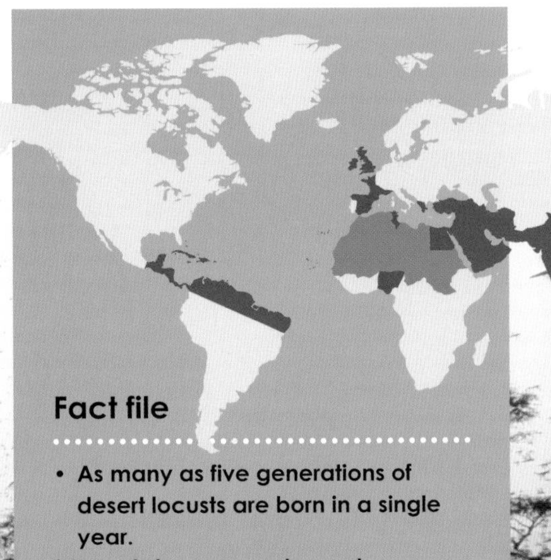

Fact file

- As many as five generations of desert locusts are born in a single year.
- Locusts lay egg pods, each containing about 100 eggs.
- When the eggs hatch, you can find as many as 5,000 babies in 1m².

59. Killer bee

Apis mellifera scutellata

It's less than 2cm long and covered in fuzz, yet this tiny insect has inspired at least five bee-grade horror movies. It's the killer bee and for the past 50 years it has been swarming northward from South America, dodging border patrols along the Rio Grande to colonise the United States.

A sting thing

This fearsome half-cousin of the humble honeybee was brought to Brazil from Africa in the 1950s. Blame Warwick Kerr, an otherwise respected entomologist and geneticist who imported the bees to try and strengthen the honeybee gene pool in local hives. His butterfingered beekeeper accidentally let 26 queen and worker bees loose, a tiny number compared with the average 40,000 bees in a single hive. Their progeny soon swarmed into Argentina, Mexico and across the water to Trinidad in the Caribbean.

The bee was first sighted in the US in 1990 and has now taken up residence in Texas, Arizona, New Mexico, Utah, Louisiana, Arkansas, Florida, California and even Las Vegas.

An estimated 1,000 people have died from the stings of killer bees since they arrived in the Americas. They are more aggressive than other bees, vigourously defending their nests and able to detect someone approaching 15m away. Once the alarm is raised, the hive will release great armies of bees, which may pursue the intruder for nearly 2km.

Killer bees spread their territories by raiding the hives of their more easy-going European

cousins, deposing the queen and replacing her with their own monarch. Then they mate with the local populace, creating hybrid bees. They're remarkably adaptable, making their homes in hollowed-out logs, heaps of rubble, cars, tree stumps, log piles and even tin cans.

These invaders harm the beekeeping industry in the Americas because the hybrids produce only a fifth of the honey of the gentler European bees – and it's more difficult and therefore more expensive to harvest. They also threaten the environment by competing with native pollinating insects.

Fact file

- The tiny town (pop. 7,000) of Hidalgo, Texas, has a statue of a killer bee. It claims to be the first place in the United States to have been attacked by killer bees and now styles itself as the 'Killer Bee Capital of the World'. It even has a hockey team named the Rio Grande Killer Bees.
- When a swarm approaches, run indoors and slam the windows and doors shut.
- Don't dive into the nearest pond to escape a swarm. The killer bees will be waiting for you to surface – and you can't hold your breath forever.

60. Yellow crazy ant

Anoplolepis gracilipes

Just because it's crazy doesn't make this ant any less dangerous. The yellow crazy ant is destroying native ecosystems everywhere from the Seychelles to Hawaii and even Australia. Most scary of all are the so called supercolonies on Christmas Island, which have destroyed or displaced as many as 20 million of the island's iconic migrating land crabs.

Christmas presence

This adaptable ant was accidentally introduced to Christmas Island during World War I. With no natural predators, it quickly established gigantic colonies, ruled by as many as 300 queens. It can expand its territory by 3m a day, and now covers a quarter of Christmas Island's forests.

The native robber, red and blue crabs barely know what hits them when a swarm enters their burrows and sprays the residents with formic acid. The blinded crabs soon die, and are eaten by the ants. Crab populations in ant-infested parts of the island have been totally wiped out.

Soil fertility declines without foraging crabs to turn over the earth. Meanwhile, the ants cause further havoc to the ecosystem by protecting the honeydew-secreting scale insects that feed on trees. More scale insects means further damage to the forest – very bad news for rainforest-nesting birds such as the endangered Abbott's booby (*Sula abbotti*), which breeds only on Christmas Island.

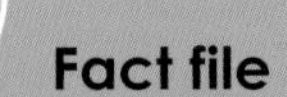

Fact file

- The ant is named for its frenzied movement and zigzagging changes of direction.
- A large species, it's also known as the long-legged ant.
- When scale insects are scarce, ant numbers fall.

Fighting back

Poisoning has only been partly successful as scientists struggle to find a toxin that will be harmless to other species but strong enough to penetrate the powerhouses of the ant's supercolonies. A massive hand-baiting campaign on Christmas Island in 2002 reduced the population by 80%. The only problem was that the colonies were expanding at more than twice the rate of baiting. Other studies are investigating methods of controlling the honeydew-producing scale insects, and the use of pheromones to cool the queens' ardour.

In Indonesia, the native Sulawesi toad (*Ingerophrynus celebensis*) has developed a taste for yellow crazy ants. It seems to be immune to the formic acid that they spray. A test by the University of Gottingen showed ant numbers were down by two-thirds on cocoa farms with toads compared with farms where fences kept the toads out. Someone with the unfortunate job of examining toad poo discovered that yellow crazy ants amounted to 75% of the toads' diet. Yum!

61. Little fire ant

Wasmannia auropunctata

The giant tortoise of the Galápagos has a small but deadly enemy in the ferocious little fire ant. This invasive alien will not only eat tortoise hatchlings but also attack the eyes and cloacae (the egg-laying orifices) of the gentle adults.

Tiny terror

The little fire ant has stormed halfway across the world, and is now considered the most dangerous ant species around the Pacific. Ideally, it prefers its landscape nicely trashed, so areas where humans have cleared the rainforest and replanted with palms, sugar cane or cocoa encourage its expansion. Much of its spread has also been through the transportation of nursery plants.

Everywhere from the Galápagos to the Solomon islands, this mini-menace has been blamed for a big drop in the population of beetles, scorpions, spiders and native ants. As well as attacking other invertebrates, it competes with them for food, usually winning.

In the wild, this ant builds nests from leaves and twigs, but in towns it invades houses and offices, particularly during flood seasons. And being a warm-weather insect hasn't stopped the little fire ant marching north. In the colder parts of the northern United States and Canada, it simply colonises heated greenhouses.

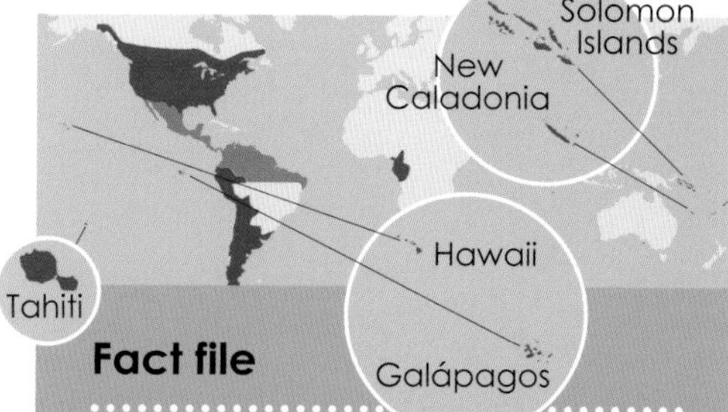

Fact file

- This species is also known as the electric ant.
- In Cameroon and Gabon it's sometimes used as a biological control, to attack cocoa-eating bugs.
- It may have spread between small islands in the Galápagos on camping equipment.

62. Big-headed ant

Pheidole megacephala

The name says it all – this ant thinks it should rule the world. Although it's only 2–3mm long, it will attack and kill young birds and other insects, causing massive biodiversity loss in its adopted home. It displaces native ant species and so reduces the number of pollinators for native plants. A study of invaded rainforest in northern Australia revealed local ants had completely disappeared and other native insects declined by 42–85%.

Protection racket

This ant is also a serious agricultural pest. Not only does it eat crop seeds, but it also protects other insects such as mealy bugs that prey on crops. The ant harvests the sweet honeydrew secreted by the bugs and, in return, protects them from spiders and other larger predators.

The big-headed ant has voyaged around the world since the 18th century on sailing ships, and now regularly hitchhikes on long-distance lorries, particularly trucks serving the plant nursery trade. It's also a nuisance for humans, chewing through electric and telephone cabling and destroying plastic irrigation hoses.

Fact file

- This ant is named for the disproportionately large head.
- Each nest contains several queens.
- The ant was first discovered on the island of Mauritius in the Indian Ocean.

SAVING THE EVERGLADES

▲ A breeding colony of double-crested cormorants along the Anhinga Trail

The Everglades is the largest subtropical wilderness in the United States, a UNESCO World Heritage Site and arguably one of the three most important wetlands on Earth. It's a huge watershed, welling up from Orlando's Kissimmee River then flowing through Lake Okeechobee and southward to Florida Bay. The water flows through a complex network of cypress swamps, mangrove forests and sawgrass prairies. The national park alone sprawls 6,475km2 and is the last refuge for the Florida panther (*Puma concolor coryi*), the world's coolest swamp cat.

▼ Florida panther, an endangered subspecies of the puma

Exotic pageant

However, more than a quarter of all species within the Everglades are from somewhere else. There are pythons from Burma, hissing cockroaches from Madagascar, caimans from Latin America and monitor lizards from the Nile. Feral pigs uproot endemic plants and as many as 300 cats per square kilometre are clawing through the migratory bird and native mammal species. The region is believed to house the greatest number of exotic plant species in the world, and the 1,000 or so species of alien bug cause US$1 billion worth of damage every year.

The plant protectors

There are nearly 1,400 non-native plant species growing wild in southern Florida, including more than 100 classed as Category 1 – the botanical equivalent of Public Enemy Number One. These disturb the Everglades by crowding out the native plants, changing the ecosystem and breeding with native species to produce hybrids.

Among the worst offenders in the Everglades are the melaleuca tree (*Melaleuca quinquenervia*), an Antipodean plant with a serious thirst. Another baddie is the Brazilian pepper or Florida holly (*Schinus terebinthifolius*), which smothers natives and habitats, but unfortunately has tasty red berries that birds like to eat and spread. Perhaps worst of all is the Old World climbing fern (*Lygodium microphyllum*), which blankets the ground, climbs up trees, and acts like the wick in a candle so forest fires flare into the canopy.

Fishy issues

Invasive fish use the artificial canals to colonise the Everglades freshwaters. Most of the new arrivals are able to tolerate some saltiness so have become established in the estuaries. The pet trade and fish farms are usually to blame for accidental and deliberate releases of these species. Among the worst offenders are the Mayan cichlid (*Cichlasoma urophthalmus*) and the blue tilapia (*Oreochromis aureus*). While efforts are being made to protect native species, naturalists say there will probably never be a time when the Everglades are totally free of invasive fish.

Figuring out how the Everglades became so overrun with aliens is a no-brainer. The wetlands are surrounded by resorts and cities, from Miami to the east to Naples on the west coast and south to the Keys. New homeowners brought in exotic plants for landscaping, and weird pets which outgrew their tanks and enclosures and were let loose in the wilderness.

The big cuts

Pressure on the Everglades intensified in the 1940s, when a huge canal network stretching 2,700km was cut through the Everglades as a flood defence. Water that should have been filtered through the natural wetlands was flushed straight out to sea. As more and more land was drained, the wilderness, farms and even cities began to dry up and the Everglades was on the verge of total eco-collapse.

Drastic situations call for equally drastic measures. In 2000, the American government began the largest eco-rehab project in history: the Comprehensive Everglades Restoration Plan.

Flushed with success

The US$10.9 billion restoration plan will protect more than 46,500km² of wetlands in central and southern Florida, and will take 30 years to complete. It will get rid of damaging canals and levees to ensure the Kissimmee River waters flow back through Lake Okeechobee and the Everglades, rather than straight out to sea. The project will restore the habitats of at least 60 threatened species of animal, by reviving the natural water flow for a million hectares of marshes. It will also provide clean water for people and farms.

War on aliens

As part of the restoration scheme, wildlife authorities are attempting to clear the Everglades of some of their most noxious invaders, starting with the animals. It's costing taxpayers about US$500 million each year. Non-native Amnesty Days are being held where pet owners can bring in unwanted exotic animals rather than release them into the wild. Meanwhile, hunters have been issued with special Reptiles of Concern licences to track down and kill the more dangerous invaders, such as the Burmese python (see page 22).

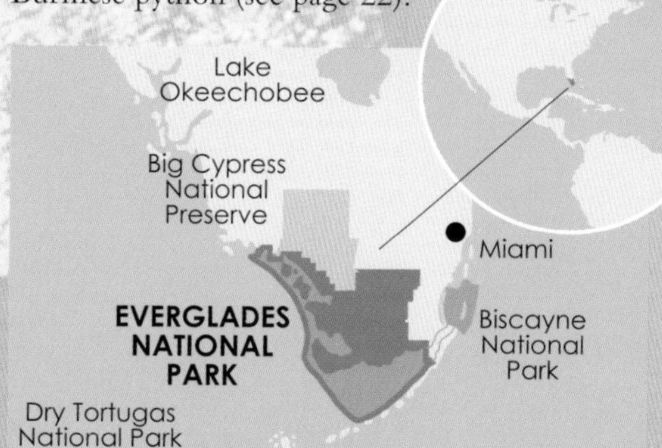

▲ American alligator

Fact file

- The Everglades is the only place on Earth where alligators and crocodiles live side by side.
- Florida has 12,500 species of native insect.
- The Everglades has 1,301 species of native plant.
- There are at least 192 species of exotic land animal and 50 species of exotic fish in Florida.

5 OUR FEATHERED FOES

Many bird species are flying into a whole lot of trouble, as bullying invaders take over new territories and force their vulnerable feathered cousins right out of the nest.

Feral pigeons in Place Mohammed V, Casablanca, Morocco

5 OUR FEATHERED FOES

If you're covered in feathers then you're definitely a bird. Lots of other animals have wings and lay eggs, and birds don't even have a monopoly on beaks. But birds have feathers all to themselves. They also have warm blood, a backbone and no teeth. Feathers enable birds to fly and steer, keep warm and attract a mate – and they're also useful camouflage. Birds probably evolved from carnivorous dinosaurs, but today there is huge variation among the 10,000 or so known species. Most fly but some, such as New Zealand's kiwis and Australian emus, stay earthbound. The class Aves contains flesh-eating raptors, expert fish-catchers, insect-eaters, strict vegetarians, and opportunists such as crows who'll eat just about anything.

▲ *Egyptian geese, native to Africa, have proved highly adaptable to the European climate.*

Flying into trouble

Birds are key components in ecosystems everywhere on Earth. They spread seeds and pollinate plants, eat crop-nibbling insects and act as unpaid garbage collectors by eating carrion.

Unfortunately, being adaptable and highly mobile means birds make particularly problematic alien invaders. Humans have aided and abetted their spread by moving them around the world, and once a bird species has been successfully introduced, it can use flight to expand its territory rapidly.

Familiar garden birds that do little harm in their home territory become serious threats to biodiversity elsewhere. Blackbirds, starlings, sparrows and finches are naturally adaptable and accustomed to living alongside humans. Taken to countries such as New Zealand by homesick early settlers, they were soon threatening or ousting native birds that were already suffering from habitat loss. These newcomers also had few predators in their adopted countries to keep their numbers naturally in check.

▲ *Ring-necked parakeets, native to southern Asia, have muscled in on garden birds in the UK.*

Attempts to use birds for the biological control of horticultural plant pests has also led to trouble. One such gamekeeper-turned-poacher is the common myna (see page 103), imported into Australia to deal with plagues of locusts, only to become a much more serious threat to many native bird species.

Cage-busters

The pet trade has also hastened the spread of invasive birds. Today, you're more likely to see a brilliantly coloured parakeet winging over a London park than a native green woodpecker (*Picus viridis*). And the breeding of apparently tropical birds, such as the African sacred ibis (see page 111), in temperate zones such as the Atlantic coast of France has been made easier by recent climate change.

Alien birds sometimes threaten biodiversity by getting too chummy with the locals. Introduced ducks breed with rarer native species and dilute the gene pool. This breeding process, called hybridisation, is especially common among wildfowl and leaves some species vulnerable to extinction. The local birds are, if you like, being loved to death.

Forced off their perches

Attacks by alien species are the biggest cause of bird extinctions in the past six centuries. At least half the bird species that have disappeared have been seen off by predators and introduced diseases such as avian flu. By far the worst-hit birds have been island species. Nine out of ten species to have become extinct in recent years have been island dwellers, unable to adapt or relocate fast enough to cope with an invasion. Beleaguered island birds usually have to deal with more than one deadly enemy. The palila (*Loxioides bailleui*), a Hawaiian honeycreeper, not only faces threats from mammals, but also the loss of its natural habitat for cattle farms, and competition for food from exotic birds and wasps. As if all that wasn't enough, avian pox from introduced mozzies means its population is now restricted to the mosquito-free highlands.

▼ Guillemots are among numerous seabirds whose food supplies are threatened by the double-whammy of overfishing and climate change.

Continental birds are suffering too. In North America, a third of the endangered birds on the Audubon Society's WatchList are on the run because of introduced species, including plants. One such exotic enemy is bufflegrass (*Cenchrus ciliaris*), which is engulfing the native plants that provide food and shelter for endangered birds, such as Costa's hummingbird (*Calypte costae*).

Other threats to birds

Climate change is putting many of the world's birds under threat. They're more sensitive than most animals to swings in temperature and wind speed, as these interfere with flight and migration habits. It is also to blame for the drying-out of salt marshes that are important habitats for wading birds, and for the loss of Arctic tundra, the breeding ground for several hundred million migratory birds. Seabirds are also going hungry as fish numbers fall in our warming oceans. Meanwhile deforestation is probably the greatest single threat to birds worldwide: many species depend upon trees, and the loss of their forests – or a reduction in the plant diversity – leaves them with no viable home.

Fact file

- **Invasive birds have a double-whammy impact by also dispersing alien plant seeds.**
- **One in ten of all bird species in the United States has been introduced, and all but a few of these aliens are causing damage.**
- **Hawaii has more introduced species of bird than anywhere else on Earth – 142 species from six continents, at the last count.**
- **Of the 17 penguin species in the world, 12 are now in decline.**
- **Today, European garden birds are spread right across the Americas, the Caribbean, southern and eastern Africa, and the South Pacific.**

63. Indian house crow

Corvus splendens

The first Gulf War brought a dangerous invader to Europe: the Indian house crow. This highly intelligent and raucous bird is believed to have hitched a ride on warships returning from the Gulf and passing through the Suez Canal. The first European sightings were in Gibraltar in March 1991, less than a month after Operation Desert Storm. Now the bird is established in the Netherlands. Its arrival in Europe is ringing warning bells, because this Asian crow has been wreaking havoc in the tropical world outside India for more than a century.

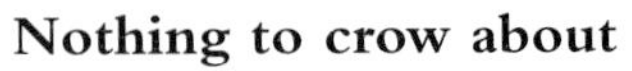

Nothing to crow about

The Indian house crow is a dangerous pest across southeast Asia and Africa, parts of the Middle East and Indian Ocean islands. It has a varied diet, devouring native lizards, frogs, small mammals and fish as well as human leftovers, and preys on the chicks, eggs and even adults of other bird species. Crow colonies will also mob any larger animal that competes with them for food.

Sadly, most introductions were made by people, often deliberately. Indian house crows were introduced to Malaysia to catch the invasive rhino beetles in the oil palm plantations. Now the crows themselves have become the pests. Their colonisation of Africa began in the 1890s, when birds were set free on Zanzibar to clear up the trash. These winged dustmen soon found easier pickings on the mainland, hitching rides on sailing ships and spreading down the east coast as far as Cape Town, South Africa.

Antisocial activities

This big, handsome crow's fondness for rubbish means it generally lives close to people and has little or no fear of humans. Loathed by farmers, it raids fruit trees, swoops upon grain crops and attacks young farm animals. In towns, its droppings are a grubby nuisance and its nests on pylons cause power cuts. Worse still, this enthusiastic scavenger can pick up and transport killer diseases.

Now communities are fighting back. In the Seychelles, a bounty was put on the head of each bird and hotlines opened so islanders could report sightings of crows and other winged invaders. Then naturalists on the Yemeni island of Socotra, a richly biodiverse UN World Heritage Site, declared war too. Socotra is home to 180 species of bird, including such endemics as the Socotra bunting (*Emberiza socotrana*). Attempts to trap the invading house crows failed, so instead schoolchildren were encouraged to collect nests containing chicks. The scheme worked.

Fact file

- **This fearless crow will dive-bomb people, snatch food from children's hands and dart through open windows to raid buildings.**
- **It has taken to nesting on mobile-phone masts.**
- **Like all crows, it is highly intelligent.**

64. Common myna

Acridotheres tristis

The ancient Greeks kept this perky bird as a pet. It was considered sacred in India for 2,000 years, and it is a great mimic, making it a popular cage bird. There's a lot to like about the cheeky common myna, yet it has fully earned its place among the world's 100 most dangerous aliens. It's a ruthless pirate, taking over nesting sites from other species and frequently preying upon eggs and chicks.

Solution becomes problem

In an all-too-familiar example of biological control gone wrong, the myna was brought into Australia to catch bugs in Melbourne's market gardens in 1862. More were imported to control cane beetles in Queensland, later in the 19th century. Now the bird is as big a nuisance to growers as the insects. It's fond of grapes, which is sour news for the wine industry, and also loves blueberries and plums. It was introduced to catch insects nibbling at sugar cane in Hawaii, too, but today is blamed for the disastrous spread of the invasive *Lantana camara* (see page 140)

Myna offences

In the Pacific, the pushy myna is becoming a serious threat to the survival of endemic birds such as the Cook Islands reed-warbler (*Acrocephalus kerearako*) and the Mangaia kingfisher (*Todiramphus ruficollaris*). On St Helena in the South Atlantic, it is one of the culprits behind a decline in numbers of the Critically Endangered wirebird (*Charadrius sanctaehelenae*).

This is a clever and highly adaptable bird. In chilly regions of New Zealand, it warms up by nesting near pig pens to use the methane central heating. In Fiji, it has taken to living along the coast and enjoys a crustacean-based diet.

The myna likes to nest in tree hollows, which poses a big problem for hole-nesting Australian parrots like the highly endangered superb parrot (*Polytelis swainsonii*). Another animal that regularly loses its home to mynas is the Australian sugar glider (*Petaurus breviceps*), a small, nocturnal marsupial that makes its home in trees. In the Indian Ocean, the myna competes aggressively for nesting sites with endemic owls and the endangered Mauritius parakeet (*Psittacula eques*).

Fact file

- **The myna is in the same bird family as the no less troublesome European starling.**
- **In India, the myna is welcome because it kills the insects that invade crops.**
- **Curried myna is considered a delicacy among northern India's Naga people.**

65. House sparrow

Passer domesticus

The chirpy sparrow is as British as the Tower of London and fish and chips. This bird has been chirruping around the British Isles since at least Saxon times and flits through the works of some of our greatest writers and historians, from the Venerable Bede to William Shakespeare. It looks innocent and charming enough, as it dust bathes in dried-out puddles and pinches crumbs from picnic tables. Yet in many parts of the world, this harmless-looking bird is regarded as one of the most dangerous aliens ever to have swooped down from the skies.

◀ *Cacti afford handy nest holes for invasive house sparrows in the southern USA.*

Your house is my house

This little brown bird may be small but it's a vicious home invader, taking over the nests and stealing the food of native species as far away as Australia, New Zealand, the Americas and the Caribbean. It survives in extreme climates, from Siberia to the tropics, and even managed to reach the Falkland Islands by hitching a ride on whaling ships.

The humble house sparrow is now the planet's most widespread bird, thanks to its remarkable ability to make itself at home wherever it travels. Before 1850 there were no house sparrows in North America. Today there are at least 150 million in the United States alone. The burgeoning US population has spread south to hook up with migrant populations from Central America, creating a sparrow superpower in the New World.

This species does well because it is fiercer and more adaptable than the local birds, and breeds for a longer season than most species. A pair of sparrows may start laying eggs during winter, so in North America it beats the native bluebirds (*Sialia* species), tree swallows (*Tachycineta bicolor*) and other hole-nesting birds to the best nesting sites. Later in the season, sparrows will raid the nests of their rivals, using their powerful little beaks to destroy eggs and chicks, and even killing the parents when they try to defend their young. In tropical Hawaii, this adaptable little thug is ousting cliff swallows (*Petrochelidon pyrrhonota*) from their rocky nests.

The disappearing cockney sparrow

Until the 1980s, the little house sparrow was one of the commonest birds in Britain and, as a familiar species, was regarded with considerable affection, especially in the cities. There were even sparrows breeding in a Yorkshire coal mine hundreds of feet below ground. Then it began to disappear.

Nobody is entirely sure why this bird that is so successful in its new colonies is dying out at home. It may be the increase in use of garden pesticides, or changes in the way we farm. But the rot may have started after World War II, when people swapped horses and carts for cars and the sparrow no longer had a steady supply of straw and grain. Some experts also believe that the decline in the number of families keeping chickens in their back gardens may also have cut off a valuable source of food for the opportunist house sparrow.

▲ House sparrows flock to feed on grain.

The impact of the house sparrow on native species is only part of the story. It also regularly threatens human interests. In India's Punjab, for example, studies estimate that one-fifth of the wheat crop is eaten by sparrows. In Russia, that figure rises to approximately one third of all the grain produced.

For such a small bird, the house sparrow is a bit of a brainbox, and persistent as well. In Australia, it has learnt to hover in front of electronic beams, in order to open automatic doors into fast-food restaurants and malls. The relentless march of the sparrow continues, with and without the help of man. It even made it to Iceland in the mid 20th century, though has yet to really prosper in this particularly challenging new territory.

Fact file

- Sparrows have been blamed for setting house fires by dropping lighted cigarette ends into thatch. (So who lit the fag in the first place?)
- A sparrow was shot for interrupting an attempt in Holland to set a world record for balancing then collapsing the largest number of dominos. The bird managed to collapse 23,000 dominos before it was stopped.
- From six million pairs in the 1970s, the British house sparrow population is now down to about three million pairs.
- Americans call it the 'English sparrow', to avoid confusion with the many native American sparrow species.
- The Communist Chinese leader Mao Tse-tung claimed that a million sparrows ate enough food for 60,000 people.

66. European starling

Sturnus vulgaris

Blame the bard for the invasion of starlings into North America. The European starling was brought from Europe as part of a whimsical scheme by New York oddball Eugene Schiffelin to introduce all birds mentioned in Shakespeare's writings – in this case *Henry IV*. Today there are about 200 million in North America, all descended from just 100 that were released into Central Park in 1890. After introduction, they soon became the villains of the piece, competing against native species and causing an estimated US$800 million worth of damage annually to crops. Not content with terrorising the US, huge flocks of starlings have since flown across the borders to Canada and Mexico.

Garrulous gangs

This strutting, glossy squawker was also introduced to New Zealand, Australia and South Africa, for insect control and for the pet trade – but with the same devastating outcome. It is great at adapting to new homes, breeding in cracks and crevices, rooftops or guttering that had previously been the nest sites of local species. It is also a master of improvisation if no obvious sites exist: one study reports a starling having nested in the wool of a live sheep.

Starlings are prolific breeders, producing up to six eggs as many as three times during the breeding season. When not raising chicks, they will settle in all sorts of habitats, from marshes to moors, along the coasts, in forests and even tundra. They go around mob-handed, descending in huge numbers on fruit or grain crops, and are just as efficient at gobbling up insects – thus depriving native insect-eaters of their food supply.

Forbade my tongue to speak of Mortimer;
But I will find him when he lies asleep,
And in his ear I'll holla – 'Mortimer!'
Nay, I'll have a starling shall be taught to speak
Nothing but 'Mortimer', and give it him,
To keep his anger still in motion.

Shakespeare, *Henry IV*, part 1

This propensity for antisocial behaviour is not confined to the countryside. Starlings roost in buildings and rip open bin bags. Their droppings can be a serious health hazard in places where mega flocks of up to 1.5 million birds have been recorded. And then there's the incessant noise, from the constant chattering and whistling, to the swish of wings when mighty flocks take to the skies.

▼ *Flock of European starlings, Texas, USA*

Fact file

- **The European starling was introduced to South Africa in the 1890s by Cecil Rhodes, who ordered 18 birds be released in Cape Town.**
- **Starlings often ride on the back of farm animals in order to eat their mites.**
- **Large flocks practise what is known as roller feeding, where birds take it in turns to forage at the front on a rotation system.**

67. Red-vented

ssage o Fiji rs d the it amoa, awaii.

BirdLife International. Today several families, known on the islands as mataqualis, are protecting 6,000ha of the forest, resisting pressure from logging companies. Deforestation and roads built for mahogany plantations create free-flying pathways for invasions by the red-vented bulbul, and other pests such as myna birds.

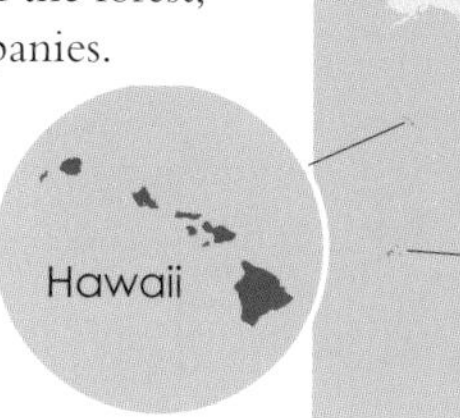

Samoa
Fiji
Tonga

Altering insects

On Oahu in Hawaii, this bulbul is now the most common insect-eater, and has cut a swathe through the population of orange monarch butterflies (*Danaus plexippus*), eating the butterflies, their caterpillars and chrysalids.

This has allowed for an explosion in numbers of the less attractive – but better camouflaged – white form of the monarch.

The red-vented bulbul has few natural predators, and as well as having a serious impact on native wildlife, it is no friend of the farmer – or the flower-grower. It causes at least US$300,000 worth of damage to *Dendrobium* orchids each year on Oahu in Hawaii.

Growers responded with bird-repellent chemicals, but the bulbul soon learnt to avoid sprayed patches. Worse, this voracious seed-eater also helps the spread of invasive plants such as *Lantana camara* and *Miconia calvescens*. Despite control efforts, red-vented bulbuls are widespread on Oahu. Now the islanders are trying to prevent the birds from establishing nesting populations in other parts of the Polynesian state.

Fact file

- **The red-vented bulbul was used as a fighting bird in India during the 19th century.**
- **Besides plants and insects, it will occasionally prey on lizards.**
- **Although it was introduced to Australia, it failed to become established there.**

68. Ring-necked parakeet

Psittacula krameri

It's a sight to make a Scotsman swear off the Scotch: a brightly coloured parakeet flitting through the woodlands in the depths of a British winter. The ring-necked parakeet, from the south Asian tropics, has been seen as far north as the Scottish Borders. This likeable but potentially dangerous alien was first spotted at large roosting in London in the late 1950s and 1960s. The pioneers probably escaped from back garden aviaries, though rumours abound of other possible origins, from film-set escapees to a deliberate release by Jimi Hendrix. Whatever the truth, there are at least 10,000 of these noisy, exotic aliens breeding in England.

Mean green breeding machine

Other countries in northwest Europe also have ring-necks, including the Netherlands and Belgium, the latter population descended from a family set free when a zoo closed in the late 1980s. They've flown along the Rhine and are spread throughout France and along the shores of the Mediterranean.

Sadly, this highly intelligent invader nests in crevices and tree holes also used by native European birds such as the house sparrow (*Passer domesticus*), European nuthatch (*Sitta europaea*) and European starling (Sturnus vulgaris). As it nests earlier in the season, it can take the best sites and food supplies from the natives.

More problem pollies

Because they are such popular pets, parrot species often escape into the wild, where they may settle down and start causing trouble. The South American monk parakeet causes problems in Florida, building bulky nests year-round in electric pylons and buildings, which short-circuit town supplies in wet weather. Scientists are also concerned that this introduced bird will spread plant diseases between trees, such as citrus cankers.

▸ A feral monk parakeet mingles with city pigeons.

Fact file

- **The ring-necked parakeet's scientific name honours the Austrian naturalist Wilhelm Heinrich Kramer.**
- **It's also known as the rose-ringed parakeet.**
- **In India it is a serious crop pest.**

Nor is the ring-necked parakeet confining its expansion efforts to Europe. It is already quite well established in the United States, and scientists are concerned about sightings in Australia. One of the most serious invasions is in the Seychelles, where the bird was first recorded in the 1970s. Ecologists are concerned that the ring-neck is competing against the endangered Seychelles black parrot, found only on Praslin, for food and nesting sites. Even worse, the invader may spread avian diseases against which its isolated island cousin has little resistance.

69. Japanese white-eye

Zosterops japonicus

There's such a thing as being too good at your job. The tiny Japanese white-eye or Mejiro was brought in to catch bugs in Hawaii gardens back in the 1920s. Now it's the most common passerine bird in the islands, and has not only cleaned up the garden pests, but is endangering native honeycreepers like the akepa (*Loxops coccineus*) and amakihi (*Loxops virens*), by eating all their food as well.

Feast or famine

A diet of mostly insects and nectar fuels this lively little bird. It's adept at expanding its territory and colonising the forests. The native birds are less skilled at finding and catching food, so are being out-competed for berries and nectar. They are also slower breeders than the introduced Japanese white-eyes. Now scientists say the akepa in particular is seriously undernourished. On one study site, scientists from Hawaii University noticed the population of akepas had dropped by two-thirds between 2000 and 2006. Not only were the endemic birds suffering from conditions associated with poor nutrition as chicks, but they were also falling victim to diseases such as avian malaria, brought in by introduced birds.

The Japanese white-eye is a tiny songbird, only about 12cm long, with a curved black bill. It goes around in flocks of up to 200 and is an agile acrobat, expert at hanging upside down where necessary to probe for larvae and insects.

Highly territorial during the nesting season, this mini feathered thief will steal material from the nests of local birds to feather its own. And to compound its crimes, it's spreading the seeds of invasive plants. In Hawaii's Volcanoes National Park, Japanese white-eyes eat the fruit of the invasive *Myrica faya*, helping to disperse its seeds. This plant is what's known as a nitrogen-fixer and it enriches the nitrogen-poor volcanic soil, which in turn helps other non-native plants become established.

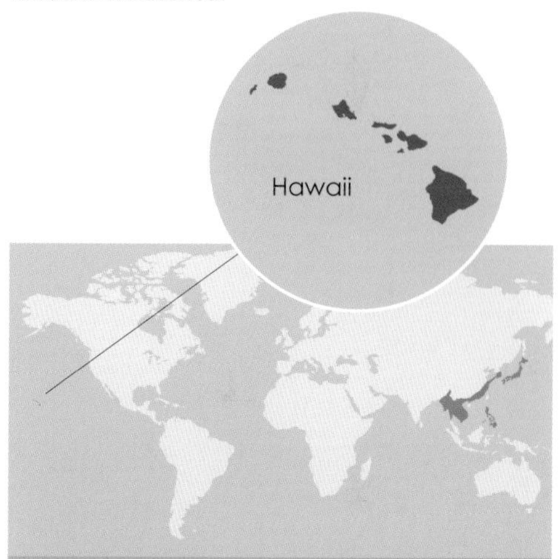
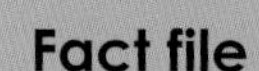

Fact file

- **The Japanese white-eye is easy to tame and is popular as a caged pet.**
- **It's often depicted in traditional Japanese art.**
- **Although generally tolerated by gardeners as a pest controller, the Japanese white-eye will damage persimmons, avocados and other tropical fruit crops.**

70. Mute swan

Cygnus olor

For a creature supposedly without a voice, the mute swan manages to make itself heard around the world. It may be much loved for its grace and elegance in Europe and Asia, but it is fast becoming highly unpopular in the United States. This big white waterbird was introduced to US lakes and gardens in the late 19th century, as an ornamental bird. Now its beauty has faded in the eyes of the Americans, as they watch the swan rather gracelessly take over the habitat of their indigenous waterfowl.

Swanning around

Mute swan numbers are burgeoning – up 20% each year in the Great Lakes. In Minnesota, the invader is chasing away the native common loon (*Gavia immer*), a handsome diving bird. In Maryland, a population of about 4,000 swans (descended from just five birds in the 1960s) is eating an alarming 4,800 tonnes of native aquatic plants every year. Meanwhile, far across the US in Oregon, the mute swan poses a threat to the survival of native tundra (*Cygnus columbianus*) and trumpeter swans (*C. buccinator*), bullying the gentler natives away from their breeding grounds.

Something has to be done. Proposals to cull the swan by the US Fish and Wildlife Service were met with anger and lawsuits by animal rights' groups (despite its unwholesome ways, the swan does have many admirers). However, in 2005 the US Department of the Interior officially declared it to be a non-native unprotected bird, even though some soft-hearted states like Connecticut still protect them. The mute swan had better keep its beak shut and its elegant head down.

Fact file

- The Queen owns all unmarked mute swans in British open water, but only lays claim to those on some stretches of the Thames and its tributaries. She shares the right to a swan supper with the Vintners' and Dyers' Companies, who were granted the treat in the 15th century.
- The annual 'Swan Upping' head count of swans on the Thames dates back to the 12th century.
- The mute swans in the Bishop's Palace at Wells, in Somerset, are trained to pull a string that will ring their lunch bell. They're NOT on the menu.

71. Black swan

Cygnus atratus

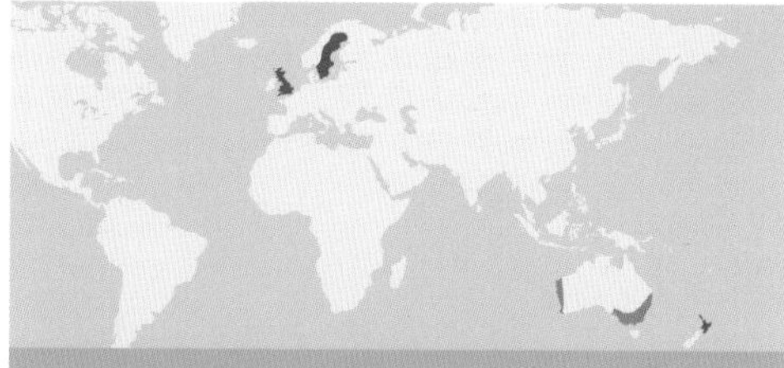

Back in the Middle Ages, the term 'black swan' was used to describe something that was impossible. To everyone's surprise, black swans were then discovered in Australia, and today the term 'black swan theory' describes a situation that was totally unexpected.

Plunderer from Down Under

Today, the black swan is having an unexpected and potentially very negative impact on British wildlife. Introduced as an ornamental bird on grand 19th century estates, it escaped when park keepers failed to re-clip its growing wing feathers. Black swans are now found in at least 170 locations around the British Isles.

They're breeding fast: the number of nesting sites trebled in a five-year study during the first decade of this century.

Black swans are aggressive and will drive other species from wetlands. They've become so well established that a black swan has been the emblem for the town of Dawlish Warren, in Devon, for more than 40 years. Hybridisation may be a future concern: there are already hybrid black x mute swans in captivity, nicknamed 'blutes'. Black swans also have the potential to damage wetland habitats.

Fact file

- A pair of black swans stay together for life.
- This bird needs 40m or more of clear water in which to take off.
- It can fly long distances but you're unlikely to see one overhead. It prefers to fly at night and rests in wetlands by day.

Taiwan

72. African sacred ibis

Threskiornis aethiopicus

Once it was a symbol of the ancient Egyptian god Thoth. Now, the African sacred ibis is more closely identified with trash. It has come down in the pecking order to be considered a serious threat to wildlife in France and Spain. In Europe it only became a problem in the 1970s when French zoos had the bright idea of breeding free-flying flocks. Naturally, they flew straight over the fence.

Holy terror

African sacred ibis are very adaptable and will eat just about anything. They've taken to roosting near rubbish dumps and slurry pits on the Loire coast, then ferrying diseases into other bird colonies. Worse, they have become predators of European animals, in particular the sandwich tern (*Thalasseus sandvicensis*) in western France. The ibises chase the terns away from their nests, then scoff the eggs. They're also varying their diets to include baby wetland birds.

In the far south of France, the ibis has been attacking the nesting sites of egrets, sometimes taking over the colonies. It also eats endangered frogs, toads and indigenous insects. Not the sort of behaviour you'd expect from a sacred bird. Ironically, the African sacred ibis is now extinct in Egypt, the country where it was once venerated.

Canary Islands

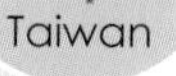

Fact file

- Ibises were often mummified and buried with pharaohs – huge numbers being bred by the ancient Egyptians for this purpose.
- Ibis fossils date back 60 million years.
- The African sacred ibis is the symbol of the British Ornithologists' Union.

73. Chicken

Gallus gallus domesticus

In the Florida Keys they don't argue about which came first: the chicken or the egg. Instead, the debate is about whether the bird known as the Key West gypsy chicken is an invasive species. Chickens have been pecking away in the soil around Florida's Key West for at least 175 years but their numbers only exploded during the 1950s when Cubans introduced a scrappier breed for the crockpot – and for cockfighting.

Scratching the surface

This new strain of chicken was more aggressive, able to fly good distances and could roost away from predators in the trees. With few predators, it prospered, spread, and began preying on native insects and lizards. It is also taking the habitat and food of native birds such as ibises and egrets, as well as other insect- and lizard-eating animals. Chickens are resourceful omnivores. In the wild, they scratch at the soil to search for seeds, insects and the small animal prey that makes up much of their diet from the first day of their lives.

On Key West, animal lovers' feathers were ruffled when an eradication scheme began. Today, the island wildlife centre runs a compromise project, where trapped gypsy chickens are rehoused elsewhere in the US as family pets.

Fact file

- The chicken is the closest known living relative (based on DNA sequencing) to the *Tyrannosaurus rex*.
- On a global scale, it outnumbers humans four to one.
- Chickens were used as oracles by the ancient Romans.
- A chicken's heart beats up to 315 times per minute.
- Each year, the poultry industry produces around 1,000 billion eggs.

Bird flu

Influenza is one of the most alarming diseases that can affect our own species. It spreads fast, is difficult to treat, and the virus that causes it can mutate into a more deadly form at the drop of a hat. Epidemics of influenza have caused massive human mortality over the centuries, and one of the more recent scares originated in a chicken farm. The viral strain HPAI A(H5N1) has adapted to be carried by birds but can infect people too. It caused an outbreak of 'bird flu' in southeast Asia in the late 1990s that soon spread to other parts of the world and reached pendemic proportions by 2008. Measures to try to contain the virus and stop its spread included massive culls of chickens in affected farms, with two hundred million birds killed. The feared global outbreak of bird flu didn't happen – this time – and there were only a few hundred human deaths worldwide.

The Hawaiian island of Kauai is overrun with feral chickens. There are various theories about their origin, one being that chickens were introduced to Hawaii by Polynesian explorers. Certainly there is evidence that they were introduced elsewhere in the Pacific, such as to Easter Island in the 12th century. However, it's more likely Kauai's nuisance birds were originally brought in for cockfights by Philippines' sugar-cane workers during the 19th century. And most Hawaiians on Kauai agree that their chicken infestation dates back to 1992, when Hurricane Iniki destroyed several chicken farms on the island, liberating hordes of cluckers. Chickens may not look exotic but they join a long list of non-native pests on Hawaii, where more species of bird have been introduced than anywhere else on Earth.

Threat to the ancestors

The domestic chicken has been flying from its coop since we began keeping it in the Far East about 10,000 years ago. It is the world's most numerous bird species, with about 24 billion worldwide. Our farmyard chicken is descended directly from the red junglefowl (*Gallus gallus*), a member of the pheasant family that is native to the foothills of the Himalayas. Modern chickens may also be partly descended from the southern Indian grey junglefowl (*Gallus sonneratii*).

▲ Sri Lankan junglefowl

The domestic chicken is able to breed with the red junglefowl, so is now threatening this species with extinction by hybridisation – the watering down of its genes. Other species in the genus *Gallus* are less threatened, particularly the Sri Lankan junglefowl (*Gallus lafayetti*), the adaptable and hardy national bird of Sri Lanka.

SOUTH ATLANTIC

The wildlife of the South Atlantic, which has evolved to deal with towering seas and Force Ten gales, goes belly-up when it meets a mouse. Along with cats, goats and invasive plants, mice have caused problems of gargantuan proportions in some of the world's most vulnerable habitats, and birds are often the unlucky victims.

Not tough enough

The South Atlantic islands stretch from St. Helena and Ascension just below the Equator, down past Tristan da Cunha, then a long way south to the Falklands and South Georgia. They're home to animals and plants found nowhere else on the planet. The remoteness of these islands means the native species have evolved in very specialised ways. They can cope with harsh climates, but not with the arrival of battle-hardened alien species, which swoop in as predators, and overwhelming competitors for scant food and shelter.

The problems caused by invasive species in the South Atlantic were recorded as far back as 1840. However, it was not until 2006 that a project led by the Royal Society for the Protection of Birds began to tackle them. With the help of botanists from Kew Gardens and local conservationists, the RSPB carried out a systematic clearance programme, which not only ousted many unwanted species, but also put in place defence systems that can swing quickly into action against any future invasions.

Rats, mice and reindeer

Tristan da Cunha is one of the most cut-off places on Earth, even though 300 resilient Tristians live in the islands. Yet mice managed to colonise Tristan's Gough Island, and began to feast on seabird chicks. RSPB scientists have been eradicating these super-rodents and making sure they don't reach nearby Nightingale and Inaccessible islands. Local fishermen help by checking their nets and boats for nibbling stowaways. Further south, on the wild and

▾ *Green turtle hatchling, Ascension*

▾ *St Helena ebony, one of many endemics on St Helena*

St Helena discovery

One of the most exciting results from the study of invasive plants in the South Atlantic has been the rediscovery of plant species that were believed extinct. The most famous of these is St Helena's *Bulbostylis neglecta*, an endemic sedge that hadn't been seen for two centuries.

Foreign animals and invasive plants have had a devastating impact in St Helena – and continue to do so. The St Helena olive (*Nesiota elliptica*) became extinct as recently as 2004. Botanists are now concentrating on clearing the aggressive white weed (*Austroeupatorium inulaefolium*) from the island. 'St Helena is the only place I've ever been where everyone uses the word "endemic" all the time,' says Kew plantsman Dr Colin Clubb. 'There is a lot of local support for the work.'

▲ Reindeer inspect an Antarctic fur seal, South Georgia

virtually unpopulated South Georgia, black rats were accidentally introduced by Norwegian whalers, and today they prey on rare seabirds like the white-chinned petrel (*Procellaria aequinoctialis*).

Hungry for venison, Norwegian whalers began introducing reindeer (*Rangifer tarandus*) to South Georgia between 1911 and 1925. Today there are two major herds of up to 2,600 animals on the island, separated by rapidly retreating glaciers. Then in the 1950s, reindeer were introduced to the Kerguelen Islands in the southern Indian Ocean. Though they do not compete with native mammals, reindeer overgraze native vegetation, destroying seabirds' nesting habitat.

The cats of Ascension

Ascension is the only breeding ground for the Ascension Island frigatebird (*Fregata aquila*), and a globally important nesting place for sooty terns (*Onychoprion fuscatus*) and brown noddies (*Anous stolidus*). Every year, green turtles (*Chelonia mydas*) swim an exhausting 2,000km from South America to lay eggs on Ascension beaches. Unfortunately, much of this natural world has been under threat from that furry killing machine, the cat. Since Napoleonic times, moggies have gobbled up 98% of Ascension's nesting seabird population – a shocking 25 million birds – forcing many species on to offshore islets.

Today, the future is brighter for Ascension's native species. Conservationists led by the RSPB began clearing feral cats in the late 1990s. By the early 21st century, Ascension was free of all but a handful of neutered pets. Noddies have returned to the main island, and conservationists hope the Ascension frigatebird will swap its offshore havens for the island that gave the buccaneer bird its name.

Falklands fight back

Another region that is successfully giving the elbow to its less desirable immigrants is the Falklands. This archipelago of some 740 islands, collectively about the size of Wales, is home to an astonishingly rich variety of wildlife – so diverse that Charles Darwin made two trips to the Falklands during the 1830s and spent twice as much time there as he did in the better-known Galápagos. Unfortunately it was ships such as Darwin's *Beagle* that were responsible for bringing rats to the Falklands. Many of the islands' native birds nest in burrows – safe from aerial attack, but not from rodents. Now the rats have been cleared from 20 offshore islands, native plants are being replanted and bird numbers are starting to rise.

▶ Breeding seabirds, such as the black-browed albatross, are vulnerable to invasive rats on the Falkland Islands.

74. Mallard

Anas platyrhynchos

The mallard is the world's most common duck, with an astonishing ability to fly as high as 6,400m (Mount Everest is only 8,848m). It migrates further than any other duck and is a rowdy and extremely gregarious bird. Unfortunately, the mallard's very enthusiasm for cosying up to the locals wherever it migrates has dealt a bitter blow to biodiversity.

Too much loving

The mallard is, along with the Muscovy duck (*Cairina moschata*), the ancestor of all modern domestic ducks. Its close association with mankind helped it to travel the world, and wherever it goes it can cause trouble. The problem is hybridisation – ie: the watering down of the gene pool to a point where a species's genetic uniqueness disappears. The mallard can breed with an astonishing 63 other species.

Now feathers are flying in South Africa about plans to cull the otherwise likeable Mallards. Some people think that hybridisation is just part of natural selection, even when local species are at risk. Mallards were shipped to South Africa between the 1940s and 1960s and conservation officials are concerned they will endanger the native yellow-billed duck (*Anas undulata*). Other African birds at risk from smitten mallards include the African black duck (*Anas sparsa*) and the Cape shoveler (*Anas smithii*).

America's favourite game bird, the black duck (*Anas rubripes*), has been blown out of the water in native areas such as New Hampshire, where the mallard is now the most common waterfowl.

Extended family

In Africa, the mallard has also cross-bred with the Egyptian goose (*Alopochen aegyptiacus*). However, mallards don't always have things their own way.

In the northwestern Hawaiian islands, migrating mallards started to breed with a long-lost relative, the extremely rare and endemic Laysan duck (*Anas laysanensis*). The hybrid ducklings were not as well adapted to blowy conditions on their isolated Pacific island and the Laysan population remains genetically strong.

Hawaii

Hybridisation

Normally, two species have to be very closely related in order to hybridise. Even if they are genetically capable of forming hybrids, often their different courtship behaviour means that the attraction simply isn't there. In wildfowl, though, the rules seem to be more relaxed.

Fact file

- **The collective noun for mallards is a 'sord'.**
- **Mallards dip just beneath the surface to feed but rarely dive deep. They also graze on land.**
- **In New Zealand, the grey duck (*Anas superciliosa*) or 'parera', is now Critically Endangered, with mallard x grey hybrids making up 95% of the country's duck population.**

75. Canada goose

Branta canadensis

In North America, the honk of migrating Canada geese overhead is a welcome sign of the change in the seasons. The bird also makes a tasty roast for Sunday lunch. Between regular hunting and natural predators such as coyotes (*Canis latrans*) and bald eagles (*Haliaeetus leucocephalus*), numbers in the Americas have been fairly well controlled. Unfortunately, this bird does not have the same population curbs in other parts of the world. In Britain, not even red foxes care to take on the aggressive Canada goose.

Gift of the goose

If it were left up to Mother Nature, only a handful of Canada geese would ever have crossed the Atlantic. Even with a good tailwind, not enough make the accidental crossing naturally to establish a population. Unfortunately, they've had a great deal of help from mankind. The Canada goose was introduced to Britain back in 1665 as a present for King Charles II. Three centuries later, its overfed descendants were culled in the park that was once part of Charles's royal gardens.

Bird strikes

Migratory Canada geese are dangerous for planes. In 1995, a US military aircraft in Alaska collided with a flock of Canada geese, resulting in the death of all 24 crew. Then in 2009, a US Airways flight lost power after hitting a flock near LaGuardia Airport in New York. Only fast and effective measures by the pilot brought the aircraft down safely in the Hudson River.

Not to be outdone, French explorer Samuel de Champlain despatched several pairs of geese to France as a gift for his monarch, Louis XIII. Successive generations of nobles and hunting groups brought in more geese, and now they are fouling the grass in 11 European countries, from Ireland to Ukraine.

These honkers nest alongside lowland riverbanks and in parks or private estates, taking habitat and food from smaller species such as coots (*Fulica atra*) and moorhens (*Gallinula chloropus*). They have also been known to chase away other adult waterfowl and kill their chicks. They eat tasty native plants and trample others underfoot. They damage crops near waterways, are suspected of passing salmonella on to farm animals, and make flooding worse by eroding the riverbanks. Canada geese also make a mess in parks and golf courses, with their slimy green droppings.

Fact file

- **The Canada goose mates for life.**
- **While it is migratory in its native North America, it tends to stay put year-round in Europe.**
- **Its introduction to New Zealand has had a disastrous impact on native waterfowl.**

76. Ruddy duck

Oxyura jamaicensis

The name of this invasive quacker sounds more like a cockney expletive than a species. The ruddy is a diving duck, attractive with its blue bill and chestnut plumage, and is common right through the Americas, from Alaska in the north to Tierra del Fuego in the far south. Unfortunately, birds introduced across the Atlantic escaped during World War II, and are now threatening the survival of Europe's native and endangered White-headed duck (*Oxyura leucocephala*).

Fence-jumpers

Some of the more puzzling species on the official British bird list are there because of ornamental wildfowl collections. Colourful ducks and elegant geese and swans look great floating across a landscaped lake, but clipped wing feathers only keep the birds in place for a certain amount of time. After the moult, the feathers regrow and if the keeper doesn't re-clip them in time, the power of flight is restored. If there are enough escapees, a free-flying and freely breeding population can develop. Established, naturalised wildfowl species in the UK include the mandarin (*Aix galericulata*), red-crested pochard (*Netta rufina*) and Egyptian goose (*Alopochen aegyptiacus*). All of these species are prospering in their new home, though none has proved as threatening to other wildfowl as the ruddy duck. Yet.

A Ruddy shame

One problem is nest raiding: ruddy ducks are quick to take over the nests of other waterfowl. However, a far greater concern is hybridisation (see mallard, page 116). The invasive ruddies are already breeding with the rare white-headed duck and watering down the genes of the European species.

▲ *White-headed duck*

The white-headed duck has only a precarious beak-hold on survival. While there are some wintering populations in the eastern Med, the only full-time European residents now are in Spain. They are fiercely protected, with numbers having risen from just 22 in 1977 to 2,600 by 2003. Hybridisation is genetic warfare and this is a battle the white-headed ducks just cannot afford to lose.

Fact file

- **Of the ruddy ducks in Europe, 95% are in the UK where the species was first introduced.**
- **The ruddy duck needs big expanses of water for take-off, and migrates in large flocks by night.**
- **It is thriving in its native Americas, with an estimated population at half a million.**

77. Feral (rock) pigeon

Columba livia

It is somewhat ironic that this bird, which helped Charles Darwin to develop his theory of evolution, should eventually become an invasive threat to wildlife on the Galápagos Islands. The rock pigeon is the ancestor of all domesticated, homing and feral pigeons, and is native to Europe where there are as many as 28 million feral birds. Now found worldwide, it is arguably the most despised bird on the planet, borrowing our balconies and windowsills as roosts to replace its natural rocky crags.

Dirty birds

Cleaning up pigeon mess from buildings and footpaths costs US$1.1 billion annually in the United States. The species carries as many as 40 diseases that are transferable to humans and other animals, such as salmonella, the avian influenza H5N1 and the stomach bug *E. coli*. In Belchertown, Massachusetts (no, really), an outbreak of *E. coli* was found to have been caused by pigeon droppings washed from the town water tower into the water supply during a storm. Pigeons' scrappy nests are also infested with ticks and lice.

The list of pigeon transgressions doesn't stop with disease. These birds steal grain and contaminate food supplies. Pigeons nesting around airports are also a danger to aircraft, causing fatal accidents when flocks are sucked into the engines.

All this makes feral pigeons costly and potentially dangerous to humans, but not necessarily lethal

to wildlife – except in the Galápagos. Feral pigeons were introduced to the Galápagos in 1972 as food for the local people. By the early 1980s they had established colonies across the islands of San Cristobal, Isabela and Santa Cruz. They carry a disease called *Trichomonas gallinae*, potentially fatal to the endemic Galápagos dove (*Zenaida galapagoensis*). Such was the threat to this unique bird that a pigeon eradication programme has been carried out across all three islands by the Galápagos National Park Service.

Darwin and the pigeons

Pigeons helped Charles Darwin develop theories that led to the publication of his *Origin of Species*. Watching pigeons snaffling oats dropped from horse feed and comparing them with fancy pigeon breeds led him to wonder whether all were descended from a common ancestor. So he set up a pigeon loft at his home in Kent and within a year had produced 15 different breeds. Today, his pigeon collection is housed at the Natural History Museum in Tring, Hertfordshire.

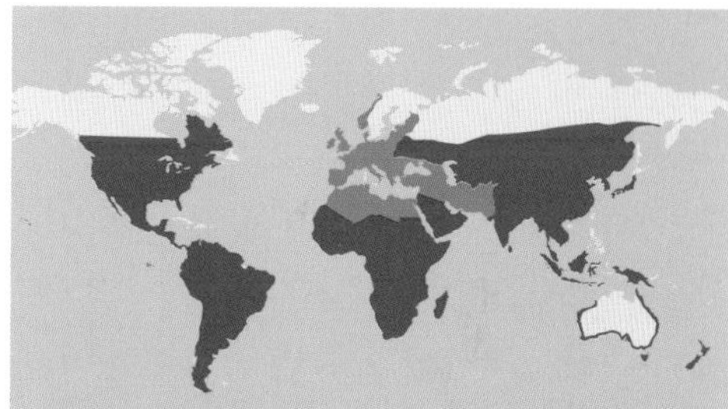

Fact file

- Baby pigeons don't leave the nest until they are fully feathered and the same size as their parents.
- Homing pigeons have been decorated heroes. One was 'Cher Ami', who received the Croix de Guerre for his military service during World War I.
- This species breeds all year round.

78. Barn owl

Tyto alba

The wise old barn owl is one of the most widespread birds in the world. There are more than 30 subspecies, living everywhere except the polar regions and isolated corners of Indonesia and the Pacific. In Europe and North America, this bird is seen as the farmers' friend because it eats so many rodents. One barn owl will typically consume 1,450 rats, mice and other small rodents in a year – more than any other animal in proportion to its size, including cats. It also has a higher metabolic rate than other owls so needs a huge amount of food just to maintain its body weight.

Silent and deadly

An assassin of darkness, the barn owl hunts by flying low and hovering soundlessly above prey. It has astonishingly powerful hearing and can locate prey by sound in total darkness. One ear is placed higher than the other, so it can pinpoint sounds more accurately. A ruff of facial feathers channels and amplifies the sound.

Sadly, this superb tracking equipment that makes the barn owl so welcome in barns has a devastating impact on populations of rare, nesting seabirds on tropical islands where there are no native owls. When an invasive barn owl can't get enough of its favourite mice on these islands, it'll turn to other birds, including those that weigh as much as itself.

Falconry fugitives

This ancient sport dates back to at least 700 BC. Many birds are shipped around the world every year to be used in the sport, raising questions as to whether or not escaped birds could become alien invasive species. This suggestion is strongly denied by falconers, who say the birds lack the skills they'd need to survive independently. Hunting birds are trained to return to the falconer and most are now fitted with tracking devices.

Owling liberties

There are alien barn owls on most islands of the Seychelles, preying on fairy terns (*Gygis alba*) and other native species such as the Seychelles sunbird (*Cinnyris dussumieri*). The owls were introduced to the Seychelles in 1949 to control rats in the coconut plantations. Now they are problematic on islands that are free or cleared of rats, and a bounty has been put on their tail feathers. The situation is particularly critical on pristine, rodent-free Aride Island. This northernmost Seychelles island is a breeding ground for 1.25 million seabirds, including the world's largest colony of lesser noddies (*Anous tenuirostris*) and the Indian Ocean's largest colony of roseate terns (*Sterna dougallii arideensis*).

▲ *Lesser noddy, Seychelles*

Barn owls are also attacking seabird colonies on Hawaii's offshore islands. One key breeding area is the Lehua Islet, home to 18 endemic seabird species. The litter beneath a barn owl roost on Lehua contained damning evidence – hundreds of seabird bones including, sadly, those of the brown noddy (*Anous stolidus*), which has now vanished from the island.

▲ *Eurasian eagle-owl*

Other invasive owls

When loggers cut down the western US forests they created a corridor for the barred owl (*Strix varia*). This alien pioneer is the latest threat to America's northern spotted owl (*Strix occidentalis caurina*), already at risk from deforestation. The larger barred owl has colonised the Rocky Mountains on both sides of the US/Canadian border. It's an extremely adaptable bird, whereas the luckless northern spotted owl is highly specialised and eats mainly flying squirrels.

In Britain there are concerns that escapee Eurasian eagle owls (*Bubo bubo*) are steadily taking over the territory of indigenous owls. Worse, the eagle owl is rumoured to prey on the endangered hen harrier (*Circus cyaneus*), although humans are more to blame for the harrier's decline. Owl groups concerned at proposals to cull the eagle owls point to evidence that they much prefer to eat rabbits, which cause an estimated £100 million damage in the UK every year. The Royal Society for the Protection of Birds says there are some fossil remains of eagle owls in Britain dating from the last Ice Age, which suggest that the bird was once indigenous to the islands.

Fact file

- **The barn owl screeches rather than hoots.**
- **It has many other names, including golden owl, white owl, monkey-faced owl, demon owl, ghost owl and screech owl.**
- **The owl was sacred to the ancient Greek goddess of wisdom Athena.**
- **Genghis Khan believed a barn owl saved his life.**
- **Owls were considered an evil omen in many parts of the world, from China to the Americas, and to this day they are feared in much of Africa.**

THE VICTORIANS

The Victorian era gave us wonderful new outside spaces, from sprawling country estates to suburban gardens, and also a great enthusiasm for all things exotic. The Victorians bequeathed us many useful garden gadgets, chief of which has to be Edwin Budding's lawnmower. They also rebelled against the gentle natural landscapes made popular by the 18th-century landscape architect Capability Brown, and transformed our countryside with technology and exotic planting.

Outdoor innovations

Victorian inventions made new ways of gardening possible in the UK. The development of cheaply manufactured glass and cast iron saw greenhouses and conservatories mushroom everywhere from Fife to Fowey. In the late 1820s, Nathaniel Bagshaw Ward noticed that plants sealed beneath glass created their own misty micro-climate. His discovery meant tropical seedlings could be shipped around the world with a better chance of survival. Soon the Victorian gentry were growing peaches and pineapples on their country estates.

▲ Rare choughs (above) are losing nest sites in Cornwall to the invasive Hottentot fig (right)

At the same time, Britain's new middle classes were moving into suburban villas, and wanting to create miniature versions of the stately home gardens. Exotics were all the rage among the wealthy landowners of the time, so it wasn't too long before monkey-puzzle trees (*Araucaria araucana*) appeared in the gardens of Kingston and Tunbridge Wells. Then came the Japanese maples (*Acer palmatum*), as Oriental gardens became all the rage. Sporting estates were also madly planting rhododendrons, as cover for game birds and for the blooms that went well with the décor of the day. That meant every border in Britain soon had a rhododendron or azalea bush.

The exotic spread

The Victorian fervour for everything foreign was peachy for the dessert tray, but brought serious problems for the natural environment. Many of Britain's worst plant invaders, such as Japanese knotweed and Himalayan balsam, were imported for country estates in the 19th century then ripped through the suburbs, jumping fences from one garden to the next.

Many of those Victorian estates in Britain are now owned by the world's oldest conservation charity, the National Trust.

Invasive species are such a concern that the Trust has launched a campaign to stamp out the alien plants on its estates by 2020. Reaching that goal won't be easy. One of the invasive species adored by the Victorian aristocracy was the Hottentot fig (*Carpobrotus acinaciformis*), which is now threatening nesting areas of the rare chough (*Pyrrhocorax pyrrhocorax*) along Cornwall's Lizard Peninsula. Getting rid of this 19th-century pest is going to involve tree surgeons abseiling down the cliffs.

Animal escapees

One of the most colourful aristocratic animal collectors was Lord Rothschild of Tring, who amassed the biggest private creature collection in the world – living and stuffed. He gave his name to a five-horned giraffe (*Giraffa camelopardalis rothschildi*), along with 17 other mammals, 58 birds, 153 insects, three fish, three spiders and a worm. He also drove a carriage with a team of zebras to Buckingham Palace, and was photographed riding a giant tortoise.

Lord Rothschild left science an incredibly valuable legacy. His collection is now part of the Natural History Museum and contains 300,000 bird skins, 200,000 eggs and more than two million butterflies. It has become an archive of DNA for many extinct species.

Unfortunately, the animal-loving lord also released exotic creatures into Tring Park, some of which escaped into the wild. One of the worst is the edible dormouse, which has become an invasive species in the Chiltern hills (see page 49). Tring and other country estates imported wallabies from Australia as lawn-mowing garden ornaments in the 19th century. They're now the only marsupials living wild in Britain, and the hardy Bennett's or red-necked wallaby from Tasmania is regularly spotted on heathland from Sussex to the wild Peak District in England, and even on the shores of Loch Lomond in Scotland.

▲ Muntjac deer

▲ Bennett's wallaby

Oh deer!

The deer park at Woburn Abbey in Bedfordshire has given generations of visitors the opportunity to watch Britain's native red deer at close quarters. Sadly, it has also been responsible for releasing exotic deer that nibble undergrowth, damaging the habitat of native birds. One of the most successful escapologists from Woburn is the little muntjac deer (*Muntiacus reevesii*), about the same size as a family dog. This deer has expanded its territory as far as Devon. Its population has soared because it can breed at any time, giving birth to two litters in a year. Another deer that staged a breakout from Woburn is the Chinese water deer (*Hydropotes inermis*). From a few strays in Victorian times, there are now about 10,000 living wild in southern and eastern England.

6 GATE-CRASHING GREENERY

We depend upon plants for our survival. Yet invasive species are pushing out shoots in every direction to destroy some of the most fragile wild habitats on Earth.

Bamboo in tropical theme garden, Lake Garda, Lombardy, Italy

We depend on plants for our survival. They provide us with oxygen, food and shelter. They form the basis of all the world's ecosystems – from deserts and dunes to rainforest, heathland, moors and mountains. If we didn't have plants, the Earth would be as desolate as Mars and unable to support life. We are inclined to take them for granted, and don't always notice when a species becomes extinct – but when an invasive plant takes over a new environment it's often only too obvious.

What makes an invasive plant?

Not all exotic plants damage their new environment. Some introduced plants, such as the potato in Europe, are useful food crops. Others provide medicine and even food or habitat for wildlife. In America, scientists say fewer than 5% of the 50,000 exotic introduced plants have become a nuisance in the wild, and only 0.3% are regarded as dangerously invasive on the scale of the kudzu vine (see page 138).

Yet the damage that tiny percentage does is disproportionate to its size. The most invasive plants spread very quickly – averaging 160km in 50 years on continents; more if they are aquatic plants and can use river currents to accelerate their spread.

An introduced plant can be invasive in some countries and not in others. The rhododendron is a highly aggressive invader in Britain, but not where it's been planted in New Zealand. And quite often the worst effects of an invasive plant are noticed far from where it was introduced. That's because wind, waterways and animals disperse the seeds. Now, climate change is also helping plants from warm parts of the world become established and invasive in temperate countries.

▲ Shrub tobacco

On the move

The spread of invasive plants to new habitats around the world is gaining momentum. It's made possible by modern machinery, the expansion of large-scale farming and forestry, and by the garden trade. Six out of ten of the most invasive plants in Britain were first planted in back gardens. Aquatic weeds are among the most difficult to remove. Pond weeds sold in garden and aquatic centres as oxygenating plants are becoming a serious nuisance in natural

▼ Foresters eradicating invading miconia, Hawaii

waterways. In Britain, these include curly pond weed (*Lagarosiphon major*) and water primrose (*Ludwigia grandiflora*).

Another big threat to wild plants is hybridisation. For example, when cultivated bluebells are introduced to parks or gardens near forests where native bluebells grow, the artificially grown species will soon take over. Now countries such as the United States are trying to put in place new codes of practice for the garden trade, asking sellers to help identify invasive plants and stop their spread. That means being more careful with shipping and disposal of soil, as well as not selling some notoriously troublesome exotics.

Invasive plants are problematic for farmers everywhere. Many introduced weeds in pastures not only compete with native animal grazing plants but are also poisonous or spiny and inedible. In the US, forage loss from invasive weeds on pastures amounts to nearly US$1 billion each year.

Advances in DNA and plant classification are helping establish an early-warning system, so new aliens can be detected and existing invasives given the heave-ho. Colin Clubbe, an expert on invasive plants at the Royal Botanic Gardens Kew, says this means botanists can distinguish between native and introduced plants in some of the world's least-known and most fragile habitats.

▲ *Invasive kudzu vine, Georgia, USA*

Stop the triffids: ways we can all help

- **Seeds are spread on boots, socks and velcro. Always clean and check, especially before visiting pristine areas such as national parks or islands.**
- **Boats and vehicle tyres spread seeds and spores. Check and scrub!**
- **Buy plants from nurseries specialising in native plants.**
- **Be careful when taking cuttings: you may be introducing unwelcome weeds to your garden.**
- **Remove invasive species from your own land, and notify authorities about any regional or local outbreaks.**

Land clearance

Clearing for agriculture, such as the felling of the world's rainforests for oil palm or coconut, creates open pathways for aggressive introduced plants. When the soil is exhausted – or the market falls away – the depleted land is ripe for invasions. Cattle grazing also allows alien scrub to take over native grassland. In Africa, that's bad news for antelope and many other native grazers.

Alien plant species are also quick to take advantage of land cleared for roads, footpaths or even firebreaks, where the natural ground cover has been destroyed and the open soil exposed. Fire is another way exotic species become established. Many successful invaders are able to regrow from their roots after a blaze, while those natives that need to grow from seed can't recover as quickly. Some invasives, such as eucalyptus, are also highly flammable and make use of fire as part of their natural regeneration.

▼ *Water hyacinth*

79. Gorse

Ulex europaeus

Yellow gorse flowers may look pretty in the hedgerows of western Europe, but elsewhere gorse is a rampant pest, threatening wildlife everywhere from Chile to Sri Lanka. Gorse was transplanted around the world by colonial settlers as an ornamental garden plant and, bizarrely, to dye Easter eggs. It was used for erosion control on newly cleared land and as animal feed. Settlers also planted it as a source of pollen for bees, not recognising local plants as equally good or better for honey production.

▼ Gorse in its native range provides important habitat for birds such as this Dartford warbler.

Fields of fire

Today, gorse is globally invasive, threatening biodiversity by replacing native species. It grows in dense clumps that cut down the grassland available to wildlife and grazing farm animals. It spreads quickly across meadows and forest edges, along the coast and around wetlands. It adapts well to varying altitudes and climates. In South America, gorse grows up to 3,500m above sea level in the Columbian Andes, rampages across the pampas of Argentina and spreads into the savannas and national parks of Brazil.

Gorse also pushes out more easy-going native plants by altering the nitrogen and acid balance in the soil to suit... you guessed it, gorse. It absorbs minerals such as calcium and magnesium to leave the ground impoverished. It's a pyromaniac's dream plant, highly flammable, with seed-pods that are opened by fire.

Some seeds are also spread naturally by birds or carried along waterways. But humans remain the gorse bush's best friend in its bid for world domination. Seeds cling to mud in hikers' boots and in velcro on clothes and backpacks. They're also spread by car, tractor and earth-moving machinery tyres. The plant produces a huge number of seeds – up to 20,000 per square metre – and they can remain dormant in the soil for up to 200 years.

Fact file

- Gorse was once used to fuel bread ovens.
- The town of Bandon in Oregon was destroyed by a gorse firestorm in the 1930s. Firefighters said the flames were too hot to be put out by water. It remains the worst fire in the state's history.
- Gorse flowers are used to make wine in Britain and to colour beer in Denmark.

80. Tamarisk

Tamarix ramosissima

This central Asian tree, also known as salt cedar, was introduced to the United States' Southwest to stop soil erosion, particularly along the river banks. Millions were planted during the Depression to stop the region turning into a dustbowl, and to give work to hungry men. Now the solution has become the problem, as tamarisk swallows up more than 400,000ha of land alongside rivers and streams, and causes the erosion and flooding that it was supposed to prevent in the first place.

Cry me a river

Tamarisk is one of the most damaging alien plants in this arid part of North America. It displaces cottonwood (*Populus deltoides*) and willows (*Salix* spp.), and deprives native birds like owls and woodpeckers of shelter. One study along the banks of the lower Colorado River showed that tamarisk supported just four species for every 40ha, compared with 154 species surviving on native trees. Tamarisk seeds contain very little protein, and are too small to be eaten by most animals. Nor are its scaly leaves palatable to birds and insects in the Americas.

This pretty and delicate-looking tree has a raging thirst and deprives native plants, animals and people of water. It also reduces the water table. Its roots bind with gravel and soil in streams to change the waterways, creating fast-moving channels that flood in spring. Meanwhile, on land, it drops highly flammable leaves that act like touchpaper to ignite brushfires. The pyromaniacal plant recovers quickly after a fire, and seizes new territory before other less flameproof plants are able to regrow.

Sodium overdose

This species gets its other name because it can grow in soils that are too salty for many species. It does this by soaking up the salty groundwater, then secreting the salt through glands in its leaves. This in turn makes the topsoil even saltier, pushing out other species of plants. It also makes the land more prone to fire.

Fact file

- **Massive amounts of seeds are dispersed in the wind and germinate fast to replace rival species.**
- **Beavers unwittingly help the tamarisk, by chomping native plants in eastern Montana and the Grand Canyon National Park in Arizona.**
- **Scientists have released a tamarisk-eating beetle into nine US western states and some national parks.**
- **China is considering planting tamarisk to stop the desert expanding.**

KEW GARDENS

Kew's Herbarium looks a bit like Hogwarts, with Dickensian wrought-iron staircases spiralling skyward, and thousands of secret drawers and cupboards. Yet this is not wizardry but cutting-edge science. The 18th-century building, in southwest London, is home to the greatest collection of plants on Earth, and it's a heavyweight resource in our effort to protect the planet's biodiversity.

▼ *Giant waterlilies*

Special specimens

To protect habitats today, we need to understand what things were like before invasive species began scrambling triffid-like across the globe. Scientists can only combat new threats if they know where species occur naturally, and when and where they've been introduced.

The centuries-old collection at Kew unlocks many secrets of the past. Some items in the Herbarium are ancient, such as the sliver of 4,000-year-old olive tree from King Tutankhamun's tomb. However, the official oldest item in the collection is a pea called *Indigofera astragalina*, plucked from the ground at Fort St George in India in 1700. The way these plants are collected has changed little since that time – specimens are dried between sheets of paper in presses, wrapped in newspaper and taken back to Kew to be stuck on to acid-free archive paper. The big difference in the 21st century is that they are also frozen to kill off any stowaway bugs.

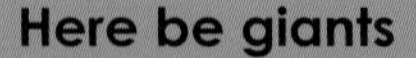

Here be giants

New species are being discovered all the time, from giant rainforest trees to rare orchids, palms and aquatic plants. Yet nearly a third are believed to be in danger of extinction. One recent dramatic discovery is the rainforest tree *Berlinia korupensis*, a 42m giant known to originate from just one small corner of Cameroon. The species was named after the Korup National Park where it was found. Kew scientists have discovered more than 100 new plant species in Cameroon since the the mid-1990s.

Not all Kew discoveries are sky-high. The smallest species found recently is a millimetre-thick wood-rotting fungus from the Kimberley region of Western Australia. Nor are all new species from distant corners of the globe. One startling find was spotted hidden among other plants in a greenhouse in Kew Gardens.

Future-proofing

Studying the DNA of specimens in the Herbarium tells scientists where plants grew at a particular time, and helps them map the

▲ Paintings of rare plants around the world by 19th-century artist Marianne North

world's vegetation. Armed with this information, they can figure out how to protect endangered habitats and help trashed ecosystems to recover. This is particularly important if we are to restore ravaged forests, and stop cleared areas being overrun with invasive weeds. It's a race against time, because the felling of tropical and temperate forests around the world accounts for a fifth of all our carbon emissions.

Intelligence gleaned from the Herbarium collection also helps us identify food plants that might be needed if one of our staple crops is struck down by bugs or disease, as happened in Ireland during the 19th century potato famine. This work is becoming more urgent as we attempt to cope with climate change, which is giving plant pests and diseases the opportunity to expand their territory.

Kew's intrepid gatherers have collected an astonishing seven million plants and fungi, with 50,000 new specimens arriving every 12 months. That makes it necessary to add a new building to the original Herbarium every half century or so.

Professor Stephen Hopper, director of the Royal Botanic Gardens at Kew, says scientists are discovering 2,000 new species every year. 'There is so much of the plant world yet to be discovered,' he explains. 'Without knowing what's out there and where it occurs, we have no scientific basis for effective conservation.'

Millennium seed bank

If the Herbarium is the HQ for plant intelligence, then Kew's Millennium Seed Bank is our mightiest weapon in our arsenal against biodiversity loss. It's the largest wild plant seed bank in the world, already protecting the seeds of 10% of the world's wild flowering plant species. More than 3.5 billion seeds from 25,000 species have been gathered. The seeds are stored at Kew and in their native countries. The good news is that none of the species collected will become extinct as long as their seeds are protected in the bank. Many seeds can survive for hundreds, even thousands of years.

▼ Seed diversity in the Millennium Seed Bank

Fact file

- **The Royal Botanic Gardens, Kew, is a UNESCO World Heritage Site.**
- **It began as a 3.6ha garden, created for George III's mother Princess Augusta in 1759. Today, it covers 121ha.**
- **Kew germinated 70,000 Amazonian rubber seeds in the 1870s, and dispatched the seedlings to establish a rubber industry in Malaysia and Sri Lanka.**
- **The Temperate House is now the largest surviving Victorian greenhouse.**
- **The cycad *Encephalartos altensteinii* in the Palm House arrived at Kew from South Africa in 1775. It's the world's oldest pot plant, and was brought to England by Captain Cook's second expedition.**
- **Kew's oldest surviving tree is a sweet chestnut (*Castanea sativa*), believed to date back to the 17th century.**

81. Wild tamarind

Leucaena leucocephala

In the 1970s, wild tamarind was hailed as a miracle tree, giving people across the tropical world a ready supply of firewood, shade and farm animal fodder. This Mexican and Central American species, also known as the white leadtree and the jumby tree, was good at stabilising sandy soil and fixing nitrogen in the ground.

▲ *A researcher in Madagascar investigates the toxic effects of wild tamarind on ring-tailed lemurs.*

Thriving with neglect

Now the honeymoon's over. While wild tamarind is a nourishing cattle food, a chemical called mimosine in the plant is toxic to many other animals. And although it's successful in reclaiming trashed land, wild tamarind is also invading rainforest and good farmland. It forms dense and impenetrable thickets. These shade native plants, and reduce the amount of food available to wild birds and insects. Today, wild tamarind has been declared an invasive weed in 20 countries, and has spread to every continent except Europe and Antarctica.

Industrialisation has encouraged wild tamarind to flourish, as farmers drift off the land to work in towns and cities. The Portuguese introduced this white-flowered tree to Taiwan in the 17th century, where it was used as fuel and livestock fodder. Regular harvesting kept the trees under control until the technology boom in the 1980s, when villagers abandoned their farms to work in factories. Wild tamarind expanded rapidly to engulf the tropical coastal forests and deserted farms of the far south, and the islands of Penghu.

Having planted the tree extensively, farmers and foresters now face the problem of getting rid of wild tamarind where it is has become an invasive weed. This is no mean feat. Wild tamarind is a very adaptable plant, able to survive in rainfall that varies widely from just 50cm to 3.5m a year. Although it started out growing along the coast and beside estuaries, it does particularly well in dry areas. Cut this tree down and it will regrow taller and stronger than before. The trees flower year-round in many parts of the world, and the seeds can remain dormant in the soil for 20 years. The seeds are also dispersed far and wide by rodents and birds, and in cattle droppings. Tamarinds flourish in dry areas prone to fire, recovering easily after a blaze and reclaiming ground before the local plants can take hold.

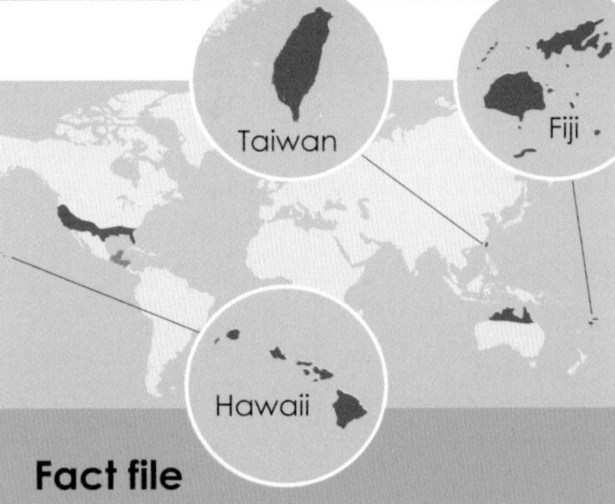

Fact file

- Unripe seeds and pods were used by native Central American people as a medicine.
- This species is being considered for use as a biofuel.
- The seeds – 'jumby beans' – are used as beads by many people in the tropics.
- In South Africa, the seed-eating beetle *Acanthoscelides macrophthalmus* has been released to try and control wild tamarind.

82. Japanese knotweed

Fallopia japonica

Japanese knotweed is an ornamental delight in a Zen garden. It was so admired by 19th-century botanists that they took cuttings back to the botanical gardens in Europe, North America and Australasia. This was disastrous. In its home territory, Japanese knotweed is naturally kept in check by more than 240 species of weed-munching insects and fungi. Yet the lush plant, with its attractive heart-shaped leaves, has few predators outside the Far East, and has become one of the most aggressive alien species on Earth.

Power plant

This pernicious invader competes vigorously against native species, with a devastating effect

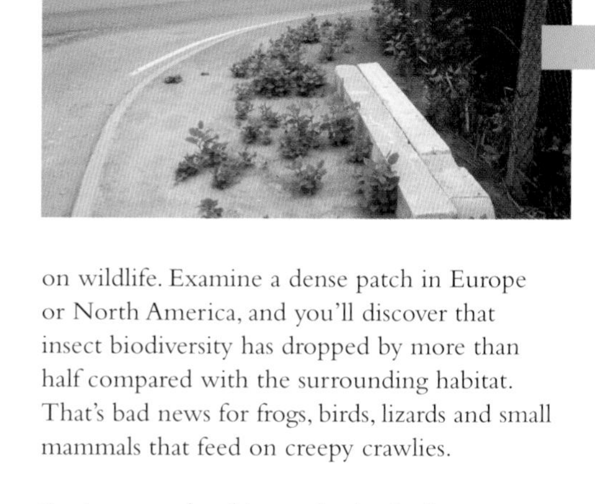

on wildlife. Examine a dense patch in Europe or North America, and you'll discover that insect biodiversity has dropped by more than half compared with the surrounding habitat. That's bad news for frogs, birds, lizards and small mammals that feed on creepy crawlies.

Engineers and architects also loathe Japanese knotweed, which can push through a metre's depth of tarmac or concrete to break up roads, undermine foundations and invade drains. It also causes flood damage by invading riverbanks, only to die back in autumn and leave the banks vulnerable to erosion.

The plant spreads at a rate of a metre a month above ground and 7m below, pushing up shoots called rhizomes from beneath the soil. It's a costly enemy: so far, Britain has spent an estimated £1.5 billion trying to control this invasive pest; Germany spends an average 30 million euros every year doing the same.

A knotty problem

Britain's fighting back with one of Japanese knotweed's oldest and deadliest enemies, a small Japanese bug known as *Aphalara itadori*. Tests by British scientists showed that this 2mm plant predator will attack the knotweed but leaves related native species alone. Trials began at isolated test sites in 2010, but it will be some years before horticulturists decide whether or not this biocontrol project is a success. Meanwhile, other scientists are concerned that we may just be releasing another potentially invasive species into the British countryside.

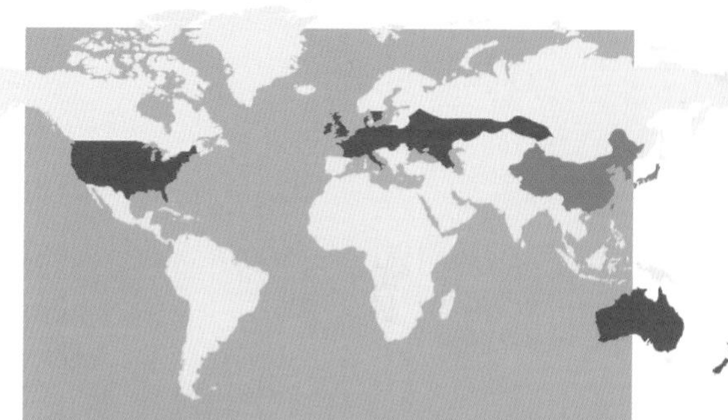

Fact file

- The plant is used in Japanese medicine and called *itadori*, which means 'take away pain'.
- It can take three years to completely destroy a patch of knotweed with weedkiller.
- Knotweed leaf litter is low in nutrients, and where it falls into rivers it may damage the riverine ecology.

83. Golden bamboo

Phyllostachys aurea

Bamboos are the fastest-growing plants on Earth. Some varieties can grow a metre in a single day, and reach full height in just three months. In prehistoric times there were vast forests of bamboo, towering more than 80m! Bamboos are among the most useful plants on our planet, widely exploited in their native east and southeast Asia for building, making furniture, cooking utensils, paper, fuel, food and medicines. Yet these perennial grasses can cause huge environmental problems when replanted in other parts of the world.

Good versus bad bamboo

Basically, although there are about 1,500 species of bamboo, they fall into just two broad categories. The first are called clumping or sympodial bamboos and stay more or less where they are planted, so are not invasive.

It's the so-called running or monopodial bamboos that cause the problems. These put out long underground stems called rhizomes, from which sprout dozens of shoots. They are incredibly difficult to eradicate and soon become dangerous invaders in their adopted territory. Running bamboos are aggressive colonisers, crowding out native plants and reducing the amount of food available to insects, birds and mammals. They are particularly adept at taking over recovering secondary forests.

One of the most common offenders is golden bamboo, declared a noxious weed in countries as far apart as Australia and the United States. It's a perennial that can grow to about 12m high in a wide range of conditions, from sea level up to 4,200m. It was introduced to the US as an ornamental grass, and to make fishing poles, back in the 1880s. It's still called 'fish pole plant' in many southern US states. Today, it's an invasive weed, everywhere from Georgia and Texas in the south, to Oregon in the West and New Jersey in the north.

Panning the gold

Golden bamboo is threatening biodiversity in Australia's Blue Mountains, a UN World Heritage Area near Sydney in New South Wales. This 1.03-million-hectare region is covered in temperate eucalyptus forest, and is home to unique plants and animals. Today, golden bamboo – planted for decades to make garden fences and windbreaks – is competing against many native plants. Authorities are trying to encourage people to replace bamboos with native plants such as banksias, hakeas and sheoaks, which provide food and habitat for endemic animals.

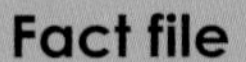

Fact file

- **An estimated one billion people in the world live in bamboo houses.**
- **Bamboo is the main food for the rare giant panda and red panda.**
- **It is a symbol of longevity in Chinese mythology.**
- **Many bamboos only flower once every 60 to 120 years.**

84. Tree tobacco

Solanum mauritianum

This plant belongs to the nightshade family, but is more closely related to tomatoes and potatoes than it is to true tobacco. It is a highly invasive weed. As a shrub or small tree, it grows up to 10m tall and crowds out native plants. It can live up to 30 years in its adopted territories and, what's more, is highly poisonous to humans, causing skin irritation and nausea if touched, respiratory problems if dust from the branches is inhaled, and death if any part of it is eaten.

Dropping in

Birds eat this plant's yellow berries, and disperse the seeds in their droppings. In this way, tree tobacco has spread across Hawaii and become established on hillsides and ridges. In South Africa it now grows in wetlands, forests and farmland, particularly in the wetter eastern side of the country. Its berries also encourage grape crop-devouring fruit flies.

The South Africans have responded with widespread eradication schemes using herbicides. They've also tried biological controls, and ring barking, with limited success. Eradication is, as always, proving more difficult and expensive than introduction. Put that in your pipe – but don't make the mistake of smoking it.

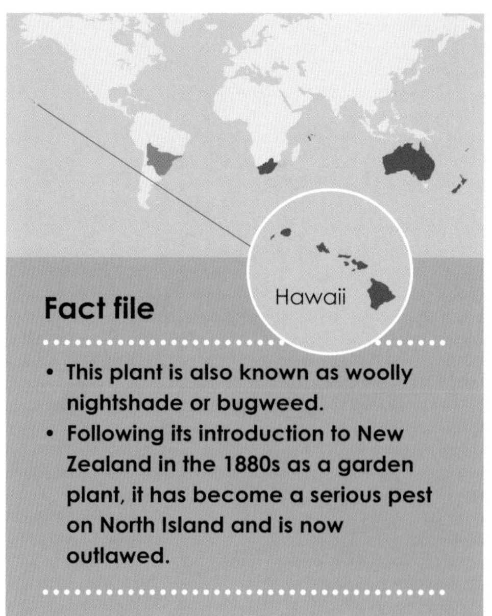

Fact file

- **This plant is also known as woolly nightshade or bugweed.**
- **Following its introduction to New Zealand in the 1880s as a garden plant, it has become a serious pest on North Island and is now outlawed.**

---

85. Wild ginger

Hedychium gardnerianum

It began as a fragrant addition to tropical gardens, but wild ginger has become a vigorous weed, threatening forests and wildlife across the world. The plant comes from the Himalayan slopes, where it has plenty of natural predators to keep it in check. Now it is choking the rainforests of Hawaii, New Zealand and the Caribbean. It is also threatening the habitat of rare insects in Brazil and endangered Azores bullfinches in the Atlantic.

Scary spice

The root looks and smells very much like edible ginger, and was used as a substitute during wartime rationing. It's an adaptable plant, flourishing in warm, wet climates, growing robustly along the coast yet able to resist cooler temperatures at 1,700m above sea level. It prefers sunlight, but will also rapidly take over rainforest.

Fact file

- **Wild ginger is banned from sale in New Zealand.**
- **It competes very effectively against native plants in almost any new territory.**

In the forest, wild ginger forms large clusters that smother seedlings underneath, disastrous for forest regrowth in sensitive areas such as the Hawaii Volcanoes National Park. It depletes soil nitrogen, and grows densely and destructively on riverbanks. This species has few natural predators away from home, and birds disperse its many seeds. You can pull the plants out by hand, but the slightest piece of root will resprout new life.

GARDENING FOR BIODIVERSITY

We will fight them in the borders, in the trenches and the plots. We will defend our territory not with guns, but with spades, hoes and trowels (well, maybe the occasional squirt gun too). Gardeners are at the front line of the war against invasive plants. If we can halt the invaders on our own patches, we can stop them sending out shoots and roots to take over the natural world beyond the garden fence.

▲ *Rowans and other berry-bearing plants can provide a garden feast for visiting birds, such as this waxwing.*

▲ *Butterflies such as this small tortoiseshell find abundant nectar in thistles and other garden 'weeds'.*

Go native

Wherever you live, the best thing you can do to protect the local environment is to go native. Find out what species grow naturally in your area, and order them from reputable garden centres such as those run by Britain's National Trust estates or the Royal Horticultural Society. Seed suppliers that deal in wildlife-friendly natives are also a good source of plants. Never collect native species from the wild – it's removing animals' food and shelter, and leaving a gap where invasives can muscle in. Planting your garden with nectar-rich native wild flowers will bring its own rewards by attracting bees, butterflies and birds to your garden. As a bonus, native plants evolved to cope with local conditions, so they usually need less watering and fewer pesticides.

Avoid using any plants known to be invasive, even if you need fast ground or fence cover. A quick-fix solution with introduced plants generally becomes a long-term garden problem when rampant stems take over. Even if non-native plants can be kept in check in your own garden, they will be difficult and sometimes impossible to control if they escape into the wild. Pruning and weeding out invasives is difficult in nature reserves or national parks, where they grow alongside fragile native species.

NIMBY (Not In My Back Yard) gardening

Reach for your trowel as soon as you spot an invasive plant. If you wait for a few weeks, the chances are it'll have taken over the border. Try to get rid of every bit of the plant, but avoid breaking it into smaller pieces that can take root elsewhere. You won't always get everything, so check for regrowth the next season, and stamp out the new plants wherever they appear. Sometimes clearing invasive weeds takes several years, but it gets easier when you learn to identify the repeat offenders. Disposing of the invaders is as important as digging 'em up. Never compost flowers – they contain seeds, so burning them is a safer means of ensuring they don't reproduce. Using herbicides is another option, but may harm other plants and wildlife.

Plant natives as soon as you remove invasives, or other aliens will move in to colonise the disturbed soil. This is also really important if you're clearing invasive species from a steep slope, otherwise you may have a problem with erosion.

Voting with your hoe

- Ask your local nursery or aquatic garden centre to stop selling plants known to be invasive. Shop somewhere elsewhere if they refuse.
- Check plant labels and read up about unusual plants before you buy.
- Quarantine new plants in your garden shed for a few weeks to check what else is growing in the soil.
- Be careful not to plant species that will hybridise with natives, such as exotic bluebells near a natural bluebell wood.
- Join a volunteer group to remove invasives from the neighbourhood or local wildlife reserves.
- Learn to recognise plant enemies and report sightings to the local council. Early detection might stop the spread.

Stop the spread

Make sure you're not spreading seeds around. On open garden days, give your shoes a good wipe between gardens and before you go back into your own pristine, natural environment. Always scrub boots and camping equipment when you return home from a outdoorsy holiday. Stop invasive grasses spreading by rinsing clippings from your lawnmower before taking it to the weekend cottage or lending it to your neighbour.

Pond life

Invasive aquatic plants kill off native species, cause flooding, and block rivers and canals for boats and anglers. So don't let the slime balls leave the pond. Even the tiniest scraps of pond plant can spread quickly in the wild. Never throw unwanted plants into the wild. Instead, let cleared plants dry out then burn the waste. Leaving them beside your pond for a few hours will give hangers-on like newts and dragonfly nymphs the chance to wriggle back into the water.

The best time of year to clear invasives from your garden pond is autumn, to cause the least disruption to wildlife. Be careful when you empty waste water – it can contain minute segments of unwanted plants. Never empty pond water into lakes, rivers, canals or even mains drains. Instead, use the water to irrigate your lawn or borders, and save yourself a few hours with the watering can.

▼ A buff-tailed bumblebee feeds on purple betony in an English garden.

86. Kudzu vine

Pueraria montana

It's the 'Vine that Ate the South'. Kudzu is a highly aggressive, climbing perennial. It was brought into the US from Asia for the 1876 Philadelphia Centennial Exposition, as a forage crop for animals and as an ornamental plant. In the 1930s, kudzu was planted across the southern states by Franklin D Roosevelt's Civilian Conservation Corps. It was used to control erosion, but unfortunately also became one of America's most destructive invaders.

Crawl for victory

An individual kudzu plant moves across country at a rate of 26cm per day – cumulatively the species is adding a staggering 61,000ha to its range every year. It smothers every native plant in its way, endangering many bird- and insect-friendly species. Kudzu vines can even uproot giant trees: this creeping monster has overcome about 810,000ha of America's eastern temperate hardwood forests.

Kudzu has smothered an estimated three million hectares of the US in total, and costs about US$50 million every year to control. Worryingly, it's since crossed the border into Canada's Ontario province, with vines having been identified on the shores of Lake Erie. The unstoppable stems also damage property, growing over fences, houses and road signs, and even causing power cuts by knocking over poles and downing power lines.

Hardly surprisingly, the US Department of Agriculture stopped promoting kudzu as far back as the early 1950s and it was declared an invasive pest in 1972. Decades later, the problem has still not gone away. Kudzu is very difficult to eradicate because half the plant is below ground. Getting rid of it requires repetitive doses of herbicides for up to ten years.

War crimes

Nor has the spread of this Asian alien been confined to the North American mainland. US armed forces introduced kudzu to the Pacific islands of Vanuatu and Fiji during World War II, as a camouflage plant. Today it is an invasive weed in both countries. It has been imported into New Zealand despite strict bio controls. The Australians are also fighting kudzu vine as it marches across the eastern corner of Queensland and into New South Wales. Without urgent action, kudzu may well become the 'Vine that Ate the Planet'.

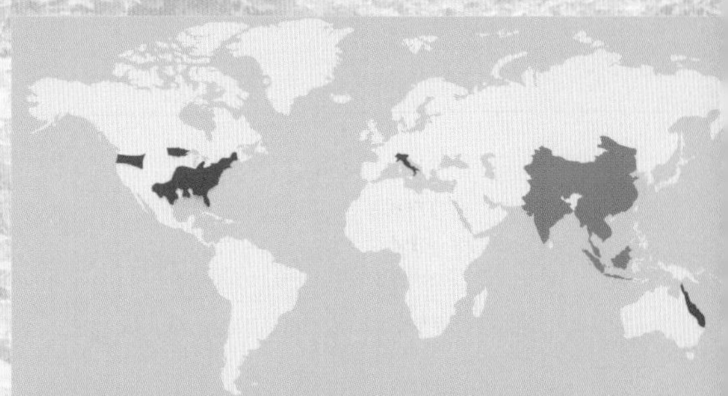

Fact file

- Kudzu is an effective mosquito repellent.
- It's traditionally used as a hangover remedy in China, while tests on rodents suggest it could be used to suppress alcohol cravings.
- Scientists believe kudzu may cause further environmental damage by producing ground-level ozone.
- Goats are used to control kudzu in Mississippi. However, they are indiscriminate foragers and will also eat endangered native plants.

87. Prickly pear

Opuntia stricta and *Opuntia vulgaris*

Native American prickly pears sailed to Australia with the First Fleet. At the same time, they sallied into the history books as the first and worst weeds ever imported into the newly settled country. Captain Arthur Philip brought several plants into the country as fodder for cochineal beetles, which the British planned to use to make dye for the British army's red coats. The first species was *Opuntia vulgaris*, which you can still see along the coast of New South Wales today. It's officially a noxious weed but its stronger cousin, the so-called common prickly pear (*Opuntia stricta*), was the one that really caused problems for Australian farmers.

Going pear-shaped

Introduced in the 19th century, this invader quickly overran the new farming country, galloping at a rate of half a million hectares every year across New South Wales and Queensland. Within a short time, 25 million hectares of land were blanketed by these 2m-high, dense and extremely prickly weeds.

Something had to be done, and fast. The answer came in the form of the appropriately named cactoblastis caterpillar (*Cactoblastis cactorum*) in the 1920s. Within six years farmers were able to move back on their land and scientists hailed the project as a runaway success. But unfortunately they spoke too soon. The caterpillars only munched prickly pears in the warmer parts of New South Wales and Queensland; they didn't like the snowy and more southerly regions. There are regular outbreaks of prickly pear today, as far south as Adelaide and across to the southern farms of Western Australia. Today, the plant is being kept under control by a combination of chemicals, cochineal beetles and caterpillars. But authorities are always watchful of this old enemy.

African adventure

Across the sea, the prickly pear has invaded and formed dense infestations over large areas of the Kruger National Park in South Africa. First discovered in the park in the 1950s, it soon covered 16 million hectares. Once again, the cactoblastis caterpillar was set free, but unfortunately its fondness for the plant is exceeded by that of elephants and baboons, which unwittingly spread the dangerous alien over a huge area.

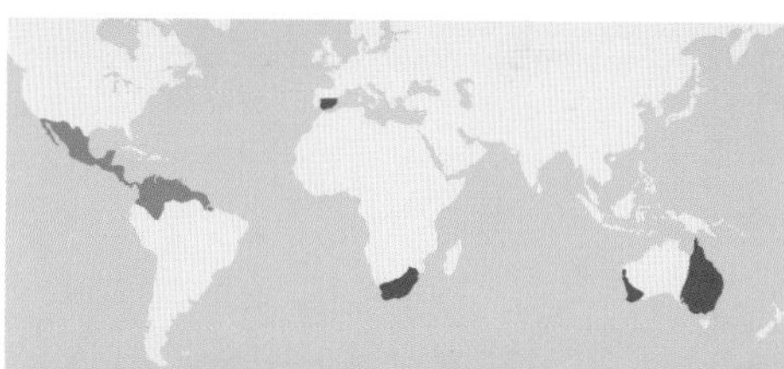

Fact file

- Mexicans have been distilling prickly pears for thousands of years, to make a drink called colonche.
- The fleshy pads on the prickly pear that look like leaves are branches that store water.

88. Lantana

Lantana camara

Lantana is also known as wild sage. But there's nothing sagacious about transplanting this weed outside its home territory. It's one of the most invasive plants on Earth, taking over fragile habitats throughout the tropical world. It isn't even a real sage: the herb mentioned in Simon and Garfunkel's song 'Scarborough Fair' is a member of the family Lamiaceae, whereas this invasive alien belongs to the family Verbenaceae.

Blame the gardeners

Lantana, native to the American tropics, was introduced around the world as an ornamental garden plant. It soon spread beyond the herbaceous borders, to become a serious threat to the survival of native plants and animals as far afield as Madagascar, Australia and Southeast Asia. Today there are some 650 varieties of *Lantana camara*, spread across at least 60 different countries and island groups.

In the Galápagos, this plant creates dense bushes that smother native plants and engulf the nesting grounds of native seabirds. In Kenya, it is blamed for competing against African plants that are the food supply for the endangered sable antelope (*Hippotragus niger*).

Perhaps worst of all for the health of the planet, lantana is adept at taking over areas on the fringes of tropical rainforests, slowing down the natural regeneration of these forests by up to three decades. In southeast Asia, it is out-competing the slow-growing native shala tree (*Shorea robust*), sacred to Buddhists and Hindus. Lantana is resistant to fire, quickly moving in to colonise burnt-out areas before the native plants have time to recover.

Partners in crime

Lantana is aided and abetted by other invasive species as it colonises new territories. In Hawaii, the seeds are shunned by native birds but spread by another alien, the common myna. Rats are also fond of wild sage berries and help to spread the seeds across tropical islands. Dense wild sage clumps also serve as havens for dangerous pests, such as malaria mosquitoes in India, and tsetse flies in Rwanda, Tanzania, Uganda and Kenya.

Making amends

Although it's considered a noxious weed in Australia, lantana has provided temporary shelter in deforested regions for the native *Exoneura* bees, and birds such as the rare black-breasted buttonquail (*Turnix melanogaster*).

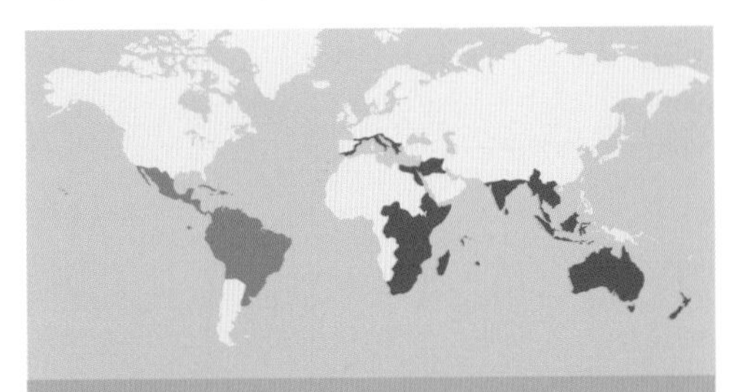

Fact file

- Lantana is being investigated for use as a possible weedkiller against another dangerous alien, the water hyacinth (see opposite).
- It has been used in herbal medicine to treat leprosy, tetanus and cancer.
- Because of its colourful flowerheads, it is also known as Spanish flag.

89. Water hyacinth

Eichhornia crassipes

In the 120 years since it was first spotted growing along Egypt's River Nile, water hyacinth has become the world's most invasive aquatic weed. This species is a valuable part of the ecosystem in its native Amazon and Pantanal wetlands, in South America. However, it has spread to clog the waterways of more than 50 countries across five continents. It chokes the life out of native plants and wildlife, is a major health hazard, and an economic disaster.

▲ *Hippos in water hyacinth, Masai Mara, Kenya*

Sink and swim

Water hyacinth spreads astonishingly quickly. Within ten years of the first sighting in Lake Victoria, it had covered 12,000ha of the lake, blocking entry to fishing ports in Kenya, Tanzania and Uganda.

The pretty blue flowers release hundreds of seeds into the water. These sink and can remain dormant for 20 years, making long-term eradication extremely difficult. Once the seed germinates, the plant is in flower within about ten weeks. It floats freely, making use of the current to drift downstream and invade new territory. Water hyacinth starves the water of oxygen and pollutes fish nursery areas. It also cuts out sunlight to native water plants, soon becoming the dominant plant species.

The human cost

Besides the problem with the natural environment, water hyacinth causes health and economic problems in some of the world's poorer countries. It slows down river flow, leading to the contamination of drinking water supplies, and creates ideal conditions for malaria mosquitoes and the snails that are hosts to the killer disease bilharzia. Hyacinth mats completely block rivers and canals, making them virtually impassable for small boats. The plants can also get wrapped around hydroelectric machinery, bridge supports and other manmade structures.

Miniature weapons

Scientists have unleashed two hyacinth-loving weevils (*Neochetina eichhorniae* and *Neochetina bruchi*), which are now chomping their way through the dense clumps in 20 African countries. In South Africa, the authorities have also released moths and mites against the hyacinth invasion. Not all biocontrol methods have been successful, because the results vary according to the local climate and other conditions. However, thanks to the weevils, the water hyacinth mat in Lake Victoria is down to less than 20% of its size in the mid-1990s.

Fact file

- **The root system makes up to half the plant.**
- **This species is sometimes used as duck food in Chinese farms.**
- **Fishermen and canoeists can help stop the spread of seeds by cleaning boats before they move from one river to the next.**

90. Coffee

Coffea canephora

Coffee in all its various forms is passionately loved by humans the world over. However, elephants, macaques and sloth bears may unwittingly be turning the cultivated low-grade robusta coffee (*Coffea canephora*) into an invasive species. Seeds of this coffee species, originally from Africa, are being spread by the animals into the Western Ghats rainforest in India, one of the richest areas of biodiversity in the whole of tropical Asia.

The threat to your morning cuppa

Sip that cup of coffee slowly, because there may be few top-ups if the coffee berry borer beetle (*Hypothenemus hampei*) gets its way. This little menace is giving coffee producers sleepless nights. It's rapidly expanding its territory, propelled by caffeine and global warming.

The beetle, virtually unknown in the 1960s, has invaded 70 countries since the planet began to heat up. It needs a temperature of about 20°C to reproduce, and Ethiopia's coffee-growing highlands hit that somewhere in the 1980s. The warmer the climate, the more beetle eggs, and the more plants are damaged. Such is the impact, that coffee production in Ethiopia – the ancestral home of the coffee bean – is down by a third compared with pre-beetle years.

Caffeine hit

The robusta plantations run up to the edges of the forest, and studies show the plant is damaging some of the native species beneath the dense canopy. Botanists also say the robusta coffee plants offer little habitat for wildlife, compared with the world's favourite arabica coffee.

However, a bigger problem lies in the clearing of rainforest to grow coffee, and with changes in the way the crops are produced. Sustainable means of growing have been replaced in many countries by large-scale monocultures. Traditionally, coffee was grown beneath fruit and shady hardwood trees, to create an ecosystem that mimics native forest. Such plantations support plenty of wildlife. In El Salvador, shady coffee plantations still make up 60% of the country's rainforest canopy.

In Mexico, scientists found shady coffee plantations fed and sheltered 180 species of bird. This is good news for the growers, as well as the native and migratory birds, because the birds tend to feed on insects rather than on the coffee beans. However, when coffee is grown on a large scale as a monoculture without other trees, the number of birds drops by 90%. Farmers have to spend more on insect control and irrigation.

Fact file

- **Coffee plants in sunny plantations age more quickly than those grown in the shade.**
- **The genus *Coffea* contains more than 90 species.**
- **Caffeine is toxic to most plant-eating insects.**

91. Coconut

Cocos nuciferato

A dangerous invader may be seizing territory while we snooze on our sunloungers. It has been quietly expanding its range for centuries, and could be the trigger for a catastrophic breakdown of tropical ecosystems. This shady invader is the exotic coconut palm tree. It may look as though it belongs on tropical islands, but don't be fooled – it's a highly damaging alien that may disrupt the balance of nature across much of the world.

Coco channel-hopping

In the right place, the coconut palm is one of our most useful trees. Virtually all of the palm can be harvested, for everything from fuel to Bounty bars, and we get through about 60 million tonnes of coconut products every year.

Unfortunately, the coconut palm is the wrong shape for most other species. Seabirds can't nest in the crown of the tree because of its spiky, sharpish throngs. No birds means no droppings or guano around the tree. That may be good news if you're taking a nap beneath the palm, but it means there are fewer nutrients in the soil for other plants, and therefore less food and shelter for native animals.

Coconuts expand their territory well. The nuts are extremely waterproof and can safely float huge distances by sea. Coconuts collected around Norway are still fertile – although it will take centuries of global warming before this plant becomes a problem species in Scandinavia. However, most tropical coconut groves were planted in the last 200 years or so, and this mass expansion – coupled with the felling of native forests – is causing environmental chaos.

A matter of taste

A study by Stanford University on Palmyra in the Pacific showed that soil around palms was 12 times lower in nutrients than soil taken from beneath native trees. To test whether or not this drop in soil goodness had an effect on wildlife, the researchers recruited hungry volunteers from two local species, a hermit crab and a grasshopper. They were fed the leaves from native forests and from plantations where palms had seeded. The animals expressed a clear preference for leaves from the native forests, rather than those from the same plants near coconut groves.

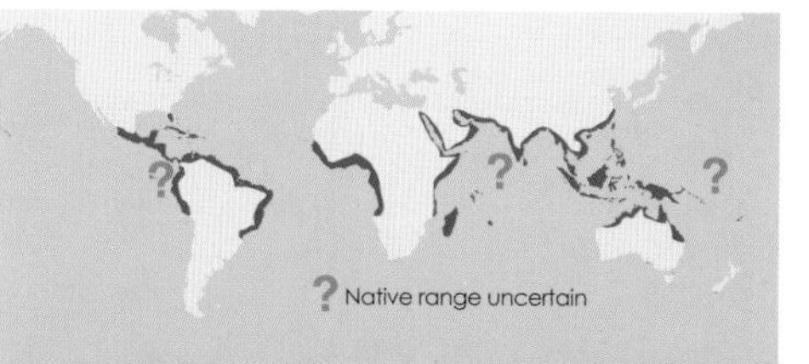

Fact file

- A coconut palm can produce up to 200 nuts a year.
- Falling coconuts kill about 150 people every year – ten times the number killed by sharks.
- A 15-million-year-old fossilised coconut was discovered in New Zealand, and named *Cocos zeylandica*.

ANTARCTICA: TIME TO CLOSE THE FREEZER DOOR

All it takes is a single grass seed attached to the sole of a boot and there goes the neighbourhood – or the ecosystem of an entire continent. Antarctica is the Earth's last mighty wilderness. It's also heating up faster than any continent on the planet. In the last half century parts of Antarctica have warmed up by 3°C, as much as five times the global average rise. That makes it ripe for colonisation by all sorts of invasive alien species.

Tentative toeholds

The invasion has already begun, with the sprouting of a blue grass, *Poa annua*, on King George Island, just 120km from the icebound continent. The seeds were probably brought to the island by a tourist bowled over by the pristine, natural environment. For about 25 million years, Antarctica was isolated by the fierce waters and ice conditions of the Southern Ocean. Now as many as 40,000 cruise liner visitors arrive every year, four times the number of scientists working across the entire continent.

The rogue blue grass isn't the first plant species to make its way so far south. Another invader called creeping bentgrass (*Agrostis stolonifers*) has already taken hold on many sub-Antarctic islands. The formula for rapid ecosystem change is simple: drop seeds from boots, velcro or elastic sock-tops into the soil, and climate change will take care of the rest.

▾ NGS Endeavour cruise ship near glacier, Antarctica

Plants in Antarctica would be seriously vulnerable to invasion by exotic species. They can take hundreds of years to grow. Only about 2% of Antarctica is free of ice, so most species are limited to a tiny habitat made up of rocks and scree slopes, or the weathered bones of whales and seals. The native plants can't adapt fast enough to compete against land-grabbing grasses from the north.

And plants are not the only dangerous aliens. Ships emptying ballast water are blamed for the arrival of the North Atlantic spider crab (*Hyas araneus*). It's a potentially catastrophic introduction, the first non-native marine invader for millions of years. Spider crabs of both sexes were discovered around the Antarctic Peninsula, the northernmost tip and most frequently visited part of the continent.

Antarctic animals

Meanwhile, scientists with the British Antarctic Survey have discovered that the South Georgian chironomid midge *Eretmoptera murphyi*, accidentally released on Signy Island near Antarctica back in the 1960s, has been steadily expanding its territory. It now outnumbers the native bugs. The larvae have been able to adapt to the colder weather on this lonely island even though they were brought in with plants that died out decades ago. The big worry is that this bug might find its way to the mainland. It may

▲ The destructive North Atlantic spider crab has invaded Antarctica via the ballast tanks of ships.

then threaten Antarctica's largest exclusively landbound animal – a 6mm flightless midge, *Belgica antarctica*.

Animals have a tough time surviving in such a harsh environment. 'I do not believe anybody on earth has a worse time than an emperor penguin,' Apsley Cherry-Garrard wrote in his story about the infamous Robert Scott Antarctic expedition, *The Worst Journey in the World*.

Species struggling to survive in such tough conditions need all the help they can get. That's why Antarctica is protected by the Antarctic Treaty, which makes it illegal to introduce exotic plants, animals and bugs. It's also against the law to disturb or harm the native plants and wildlife.

Fact file

- Antarctica is one and a half times the size of the US.
- The ice sheet covering Antarctica is up to 4.8km thick.
- It holds 70% of the world's fresh water.
- If this ice melted, the oceans would rise by more than 60m.
- Antarctica is the coldest and windiest continent on Earth – and drier than the Sahara Desert.

Why we must protect Antarctica

The key to understanding our climate is locked away in the ice cap of Antarctica. It's like a chilly library, with ice layers like pages revealing information about our changing climate for the last million years or so. The studies carried out in Antarctica may protect mankind for centuries to come. One such breakthrough was the discovery of the hole in the ozone layer above the continent, by scientists with the British Antarctic Survey. They revealed the damage done by chlorofluorocarbons (CFCs) that decompose our protective ozone layer. As a result of their studies, the world has now banned the use of the CFC chemicals that were causing the damage.

92. Japanese kelp

Undaria pinnatifida

It may be the main ingredient in miso soup, but Japanese kelp or wakame has a less appetising impact on the environment in other parts of the world. This luxuriant fleshy seaweed, native to the coasts of China, Korea and Russia as well as Japan, grows up to ten times faster than most varieties of kelp. It soon creates a dense underwater forest, shading out the light for native plants and reducing the food supply for fish.

Kelp wars

One ecosystem seriously under threat from Japanese kelp is the giant kelp (*Macrocystis*) forest surrounding the coastline of parts of South Australia, Victoria and Tasmania. This is a unique marine environment, with strands growing up to 30m tall. They provide shelter for many different species, such as rock lobsters and abalone, as well as seabirds that forage among the kelp on shore.

Japanese kelp was also discovered in San Francisco harbour, in 2009. Naturalists reacted quickly before the invasive plant could release millions of spores and colonise the coast – endangering native fish and coastal animals such as the sea otter.

Their concern was justified, because Japanese kelp spreads incredibly fast. Within seven years of having been seen around the wharves of Argentina's Puerto Madryn, it had spread at least 500km to the south.

Stowaway seaweed

Scientists monitoring the impact of Japanese kelp believe this invasive alien was introduced accidentally to coastal areas around the world, in ballast water on cargo vessels. It may also actually attach itself to the hulls of ships. It has hitchhiked to new grounds on aquaculture imports, such as when Japanese oysters were shipped to France. This underwater weed makes life difficult for fishermen, wrapping itself around mussel buoys or ropes, clogging oyster racks and slowing down the circulation of water in fish farms.

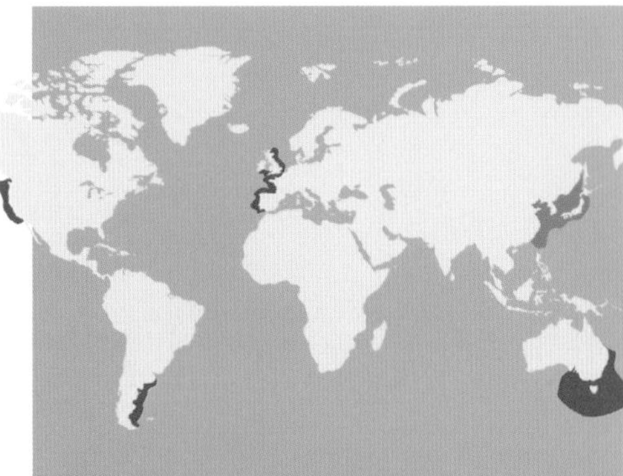

Fact file

- Kelp forests grow in nutrient-rich waters below 20˚C, as far afield as the Arctic and Antarctic circles.
- Large kelp farms may be created in the future, as a source of renewable energy.
- Eating kelp may help with weight loss, through a carotenoid called fucoxanthin.

93. Mexican thorn

Prosopis species

Mexican thorn or mesquite is a tough customer. It grows fast wherever it's planted, and tolerates very salty conditions and drought. It's extremely thirsty, with long tap roots able to reach down 20m to the water table during droughts. It's also quick to regenerate after bush fires, so is able to capture fresh ground before the local plants have a chance to recover.

Thorn again

Once established, Mexican thorn forms impenetrable clumps anywhere from 3m to 15m high, smothering other plants. It's now an invasive weed across millions of square kilometres of farmland worldwide.

Mexican thorn was planted in Australia during the 1920s for farm animal feed and is now considered one of the 20 most noxious weeds in the country. It has taken over much of rural Queensland, Western Australia and the Northern Territory. Unfortunately, farm animals and insects help spread the plant, attracted by Mexican thorn's sweet pods and nectar.

As far back as the mid-1980s, the US was spending up to US$500 million each year controlling Mexican thorn. Other clearance efforts have been attempted in Sudan, Argentina and India but the plants tend to return because the seeds can stay dormant in the ground for several years. In South Africa, scientists are now fighting back with American seed-eating beetles.

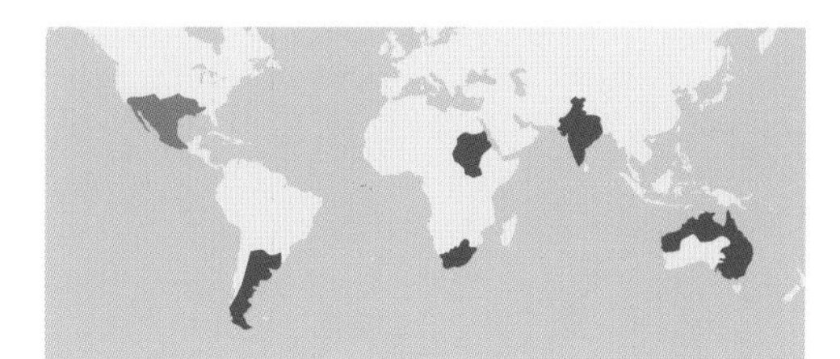

Fact file

- Some varieties are used for firewood in Indian villages.
- The wood of larger plants is used to make high-quality furniture.

94. Blackberry

Rubus discolor and *Rubus niveus*

They may taste delicious, but transplanting blackberries to vulnerable islands like the Galápagos or Hawaii leaves a bitter taste in the mouth of conservationists. In Hawaii, the so-called Himalayan blackberry (*Rubus discolor*) forms dense thickets that smother other plants. It can produce up to 13,000 seeds per square metre, which can remain dormant in the soil for several years.

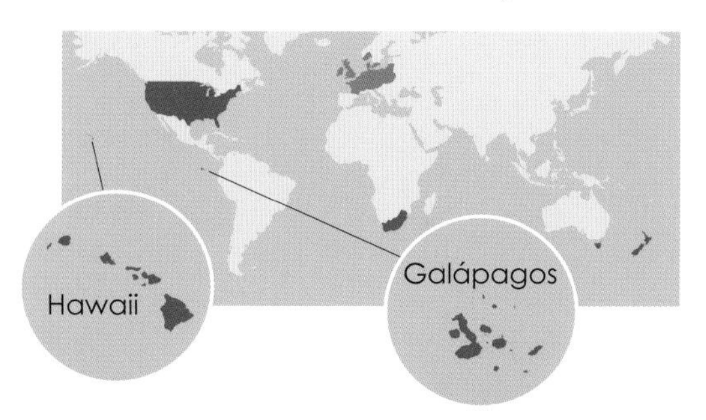

Bramble scramble

Meanwhile, introduced blackberry is one of the greatest threats to native plants in the Galápagos Islands. There are several varieties of blackberries, but the most invasive is *Rubus niveus*. It has taken over much of the islands of San Cristobel and Santa Cruz, where some of the seeds are spread by giant tortoises.

Fact file

- Blackberries are not real berries: each is a small cluster of fruits, called drupelets
- An old wives' tale warns against picking blackberries after Michaelmas (29 September). It's said that when the Devil was cast out from Heaven, he landed on a blackberry thicket and cursed the bush. So any blackberries picked after that date will be splattered with devil spit.
- Blackberries have been eaten in England since at least Neolithic times.

95. Himalayan balsam

Impatiens glandulifera

It may look pretty, but when Himalayan balsam moves into fresh ground in the British Isles it pushes out up to a third of the local native plants. This highly effective coloniser belongs to the genus *Impatiens*, so called because of the way its members rapidly expand their territory. In this case, it's by exploding pods that scatter seeds up to 7m. Each Himalayan balsam plant produces about 1,000 seeds, so it's able to spread and form dense clumps over large areas in a very short time.

Tough and sweet

This mountain invader also produces more nectar than local plant species, which seduces bumblebees and other insects into favouring it above the less generous natives. That means fewer British native plants are pollinated.

As the name suggests, it originated in the Himalayas, where tough climate conditions and local weevils and fungi limit its expansion. It only grows about a metre tall in Nepal, compared with a whopping 2.5–3m in the British Isles. Here, in its adopted home, it has become the region's tallest annual plant.

Himalayan balsam was introduced as an ornamental garden plant in 1835 and, having no natural predators, had spread into the wild across England, Wales and Scotland within 16 years. It has since made it across to Northern Ireland and the Republic of Ireland, where it has become equally invasive.

Ravaging the riverbank

This invasive plant does particularly well in fragile, moist environments such as river and stream banks, marshes and shady woodland. It does huge local damage by growing in dense clumps, which shade out local species such as young willows that are food for many native insects. It also threatens many rare plant species, such as native orchids.

Further damage is done in winter when this annual plant dies back along riverbanks. Without the traditional native grasses to hold the soil in place, the riverbanks collapse, polluting waterways and adding to the risk of flooding. Such is the risk to Britain's natural biodiversity that anyone found spreading this plant risks a fine of £5,000, or even six months in the slammer!

Fact file

- This species is sometimes known as 'kiss-me-on-the-mountain plant', or 'policeman's helmet'.
- Walkers, farmers and fishermen should take care not to spread the seeds on their boots or kit.
- The lightest touch against a ripe pod causes it to rupture violently.

96. Giant hogweed

Heracleum mantegazzianum

If you touch this plant, the chances are you'll end up queuing in a casualty ward. In just one year in Germany, about 16,000 people needed urgent medical care after coming into contact with giant hogweed. It originates from Asia, but is now one of the most invasive and harmful weeds in western Europe and North America, with a fearsome but fully justified reputation.

A pig of a plant

Giant hogweed is a phototoxic plant, which means it contains a chemical that makes your skin hyper-sensitive to sunlight. First comes the itching, then two days later the blisters appear. The scars may last for life. Worse happens if you rub the sap of this dangerous alien into your eyes. Even small amounts can cause blindness.

The tall and, admittedly handsome, white-flowered plant was brought to western Europe in the 19th century by keen amateur botanists, to be used in ornamental gardens. It's a perennial, which means that it can regrow from its roots year after year. Adapted to surviving freezing conditions in parts of Russia and Georgia, the giant hogweed has flourished in the much gentler climates of the West. It produces seeds from late spring into mid-summer and these seeds are carried widely along streams and rivers. It quickly monopolises riverbank habitats, pushing out native plants and causing the local biodiversity to fall.

This plant is such a vigorous invader and so highly toxic that it is now illegal to plant giant hogweed in many countries of the world. In 2011, a £2.6 million project was announced to try to eradicate it from Scottish and Northern Irish riverbanks. Such is its notoriety that this modern-day triffid inspired the Genesis sci-fi hit song 'Return of the Giant Hogweed', in which the plant has evolved to think for itself and is threatening the entire human species.

Fact file

- Hogweed is used as a spice in Iranian cooking.
- It's a member of the carrot family.
- Oddly, and as the name suggests, it has no ill effects for pigs. Sheep, too, can eat this towering 5m plant without suffering any unpleasant symptoms.

97. Eucalyptus

Eucalyptus spp

Firemen call eucalyptus the gasoline tree. It's one of the most flammable plants on Earth, with highly combustible oil, hanging strands of bark like touch paper and quick-drying leaf litter for kindling. Eucalyptus helps spread devastating fires in its native Australia, but the long-term effects are often worse in areas where it has been introduced, and native plants don't have the ability to recover quickly. In California, the Arastrado Preserve has still not fully regrown after a eucalyptus fire in the 1980s.

Firestarter

Eucalyptus forest regenerates fast by using what are known as epicormic shoots, which sprout from dormant buds inside the branches. There are also lignotubers in the crown – these are buds containing nutrients to help the trees regrow after a fire, when other much-needed sources of tree food are lacking.

Eucalyptus is also a thirsty tree, soaking up water through an extensive root system. Waterside groves will completely dry out what had been year-round streams. The interloper pushes out native plants by secreting toxins into the soil, and blanketing the ground with bark and leaves so other plants struggle to put down roots. Then, by soaking up available water, it denies moisture to other species.

Global domination

Eucalyptus, also known as gum trees, are among the most widely planted trees in the world, spread from Timbuktu down to South Africa, down through the Americas, the Middle East and around the Mediterranean. About 100 of the 600 species are routinely grown outside their homeland. They were exported from Australia through the 19th and 20th centuries as a fast-growing timber for pulp, fuel and shade. They expand their territory quickly, displacing other plants but offering little shelter or food to native plants and animals.

On the US west coast, a California oak will support 100 times more insects than the now commonly grown eucalyptus. In Madagascar, much of the original native forest has been replaced with eucalyptus, threatening biodiversity by isolating remaining natural areas such as Andasibe-Mantadia National Park. In Portugal and Galicia, oak forests have been also been replaced with eucalyptus pulpwood groves. Losing their natural forest is disastrous for indigenous wildlife, such as the critically endangered Iberian lynx (*Lynx pardinus*). Replacing native trees with eucalyptus also cuts down nesting opportunities for pollinating insects, particularly bees.

◀ *Brush fire in gum tree forest, Victoria, Australia*

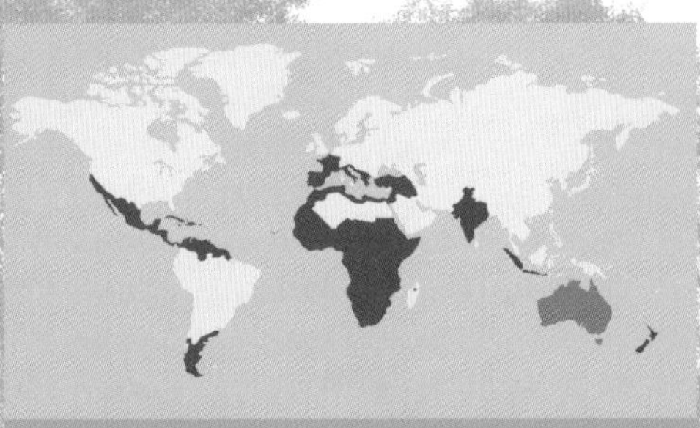

Fact file

- **Eucalyptus was introduced to the rest of the world by British botanist Joseph Banks, on the Cook expedition in 1770.**
- **Eucalyptus is considered the world's top-quality pulping species, due to its high fibre yields. It is used to produce biofuels.**
- **The oil is an effective natural mosquito repellent.**

98. Black wattle

Acacia mearnsi

If you buy a soft leather bag or belt, there's a fair chance it will have been treated with the bark of the Australian black wattle. This 5–10m tree is grown around the world for its tannin-rich bark. Its timber is used to build homes and to make woodchips. However, as useful as the black wattle may be, this evergreen is fast becoming one of the world's worst invasive plants.

Troublesome tree

The black wattle has spread quickly outside plantations and gardens and is now considered a noxious pest in countries as far afield as South Africa, Tanzania, some islands in the Indian Ocean, the United States, South America and

around the Mediterranean. It kills native plants around the South African Cape area by raising nitrate levels in the soil.

Black wattle has a powerful thirst, usually needing more water than indigenous species. It soaks up moisture, reducing the run-off into rivers and wetlands that are vital wildlife habitat.

In South Africa, botanists are letting rip with a biological control in the form of the black wattle-loving weevil *Melanterius maculatus*. Around Stellenbosch, they are also using a gall midge that prevents the plant forming its pods.

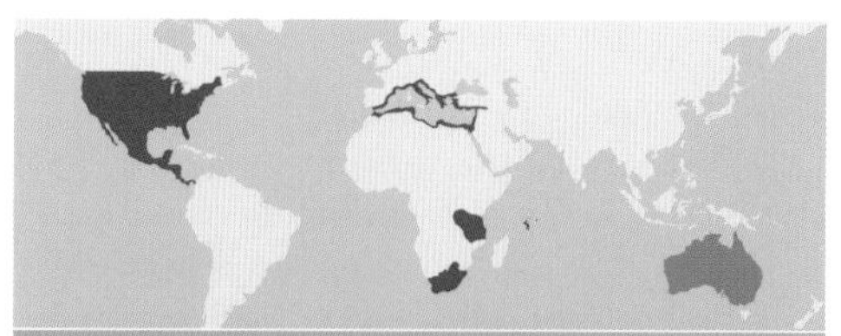

Fact file

- In Australia, the fast-growing black wattle helps the forest recover after raging fires.
- Aborigines used the bark to make antiseptic lotions for cuts and burns.

---

99. Australian pine

Casuarina equisetifolia

There's no smile on the face of the threatened American crocodile (*Crocodylus acutus*) as the invasive Australian pine threatens its nesting grounds in the Florida Everglades. Nor are the green sea turtles (*Chelonia mydas*) and loggerheads (*Caretta caretta*) fans of this Australian tree, which rapidly takes over their habitat.

Salt-seeker

This tree invades beaches, dunes and mangrove swamps, displacing native plants that provide food and shelter for endangered wildlife. It flowers constantly, scattering huge numbers of seeds to the winds and growing up to 3m per year, eventually reaching 35m tall. Its territory in Florida increased four-fold from 1993 to 2005.

Once Australian pine has established a clump, its canopy shades out native plants. It then secretes chemicals that turn the soil into a no-grow area for local vegetation. The tree also pushes out native plants that stop the beaches from eroding. It has shallow roots and often topples during storms and hurricanes, causing significant damage to property. In Florida's Everglades, this species has seriously degraded the habitat of the American crocodile and other reptiles.

Fact file

- Aboriginal people in Australia traditionally ate the resin of this tree.
- The scientific name *Casuarina* comes from the Malay word for the cassowary, because the leaves resemble its feathers.

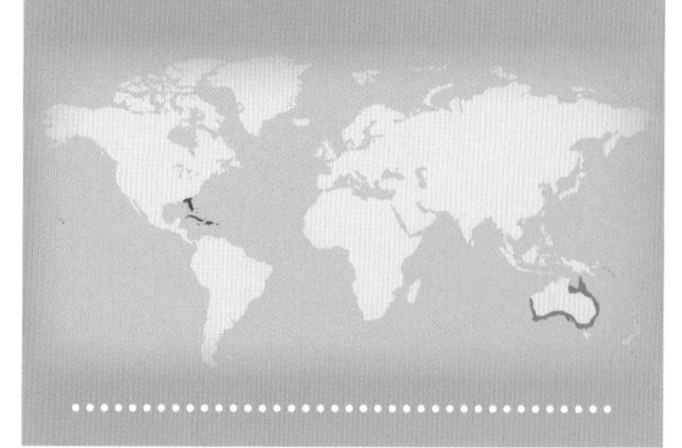

100. Rhododendron

Rhododendron ponticum

Rhododendrons were popular with Victorian country gardeners for their blowsy, colourful blooms. The species *R. ponticum* arrived in the British Isles during the 18th century, brought back by plant enthusiasts from Asia during the heyday of the British Raj. It was widely transplanted in country estates and parks, until it became apparent that *R. ponticum* needed little help to continue its spread across the countryside. It can produce millions of seeds, also spreading by putting out suckers from its roots and by layering where branches touch the ground.

Starving the woodlands

This particularly invasive species of rhododendron has become an aggressive menace across Britain and Ireland, threatening livestock and wildlife, and choking the life from some of the British Isles' ancient oak woodlands.

It forms dense and largely impenetrable clumps, standing up to 8m tall. These compete against native plant species for space and sunlight, forcing out those that provide food and shelter for insects and birds. It tolerates shade and is a particular nuisance in regions with a warm, humid Gulf Stream climate, such as the west coast of Ireland, and the Atlantic forests of Scotland and Wales.

Rhododendron also causes a big headache for farmers and estate managers, being poisonous to horses and other grazing animals. The animals eventually learn to avoid it, and concentrate on munching native shrubs, giving the plant interloper yet another advantage. Not only does it push and shade out native plants, but also drops toxic leaf litter that makes the soil acidic. This makes it difficult for other plant species to regenerate, and reduces invertebrate numbers. Studies show bird populations are generally lower in British forests overrun with rhododendrons. The invasive *R. ponticum* also carries a fungus called *Phytophthora*, which attacks native oaks.

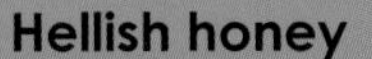

Hellish honey

R. ponticum secretes a toxin known as grayanotoxin in its pollen and nectar. This can poison people who eat honey made by bees that feed on rhododendron flowers. It's not deadly, but is liable to have a mild hallucinogenic and laxative effect – an unfortunate combination. This is nothing new: the ancient Greek scholar Xenophon wrote that soldiers in Asia Minor became ill and behaved strangely after having eaten honey in a village planted with rhododendrons.

▲ *Volunteers clear invasive rhododendron from the UK Island of Lundy.*

Fact file

- ***R. ponticum*** **is just one of more than 1,000 species of rhododendron, not all of them invasive.**
- **It's the official flower of both Indian- and Pakistan-controlled Kashmir.**
- **It would cost £25 million to clear invasive *R. ponticum* from the Loch Lomond and Trossachs National Park.**

▾ *Invasive, rhododendrons beside Loch Torridon, Scotland*

Conservation contacts

You will find further information about the species in this book from the following organisations, all of which are working in one way or another to combat the threat of invasive species around the globe.

BirdLife International
A global alliance of conservation organisations protecting birds and their habitats
www.birdlife.org
BirdLife, Wellbrook Court, Girton Road, Cambridge CB3 0NA, UK
Tel: +44 (0)1223 277 318

Durrell Wildlife Conservation Trust
Founded by the late Gerald Durrell and doing important project work with some of the world's most endangered animal species
www.durrell.org
Durrell Wildlife Conservation Trust, Les Augrès Manor, La Profonde Rue, Trinity Jersey, Channel Islands JE3 5BP
Tel: +44 (0)1534 860000

International Union for the Conservation of Nature (IUCN)
A global organisation that addresses key environmental issues worldwide and publishes the Red Data List of Threatened Species
www.iucn.org
Rue Mauverney 28, 1196 Gland, Switzerland
Tel +41 22 9990000

Invasive Species Specialist Group (ISSG)
A remarkable global group raising awareness about the threat from invasive species to ecosystems and vulnerable plants and animals
www.issg.org

National Audubon Society (Audubon)
A US-based conservation organisation looking after birds and restoring their habitats
www.audubon.org
225 Varick Street, New York, NY 10014
United States
Tel: 212 979 3000

National Trust
One of the world's oldest and most respected conservation organisations, looking after estates across Britain
www.nationaltrust.org
The National Trust, PO Box 39, Warrington WA5 7WD, UK
Tel: +44 (0)844 800 1895

Royal Botanic Gardens, Kew
The world's greatest plant collection and home to the Millennium seed bank
www.kew.org
Royal Botanic Gardens, Kew, Richmond Surrey TW9 3AB, UK
Tel: +44 (0)20 8332 5655

Royal Forest and Bird Protection Society of New Zealand (Forest and Bird)
Working to protect native biodiversity in New Zealand
www.forestandbird.org.nz
Level One, 90 Ghuznee Street, PO Box 631
Wellington 6140, New Zealand
Tel + 64 4 385 7374

Royal Society for the Protection of Birds (RSPB)
A UK-based conservation organisation whose remit goes beyond birds to protecting some of the most fragile habitats and marginal species on Earth
www.rspb.org.uk
The Lodge, Sandy, Bedfordshire
SG19 2DL, UK
Tel: +44 (0)1767 680 551

The Wildlife Trusts (TWT)
A network of regional wildlife trusts across the UK doing excellent conservation work at a local level
www.wildlifetrusts.org
National Office, The Kiln, Waterside
Mather Road, Newark, Notts NG24 1WT, UK
Tel: +44 (0)870 036 7711

Wildfowl and Wetland Trust (WWT)
Founded by Sir Peter Scott, with a particular interest in the conservation of wetland birds
www.wwt.org.uk
Wildfowl & Wetlands Trust, Slimbridge
Gloucestershire GL2 7BT, UK
Tel: +44 (0)1453 891900

Woodland Trust
The UK's leading woodland conservation charity, planting trees and fighting to protect ancient forests
www.woodlandtrust.org.uk
The Woodland Trust, Kempton Way, Grantham
NG31 6LL, UK
Tel: +44 (0)1476 581111

World Wide Fund for Nature (WWF)
A global charity protecting wildlife and nature
www.wwf.panda.org
Av. du Mont-Blanc 1196, Gland, Switzerland
Tel: +41 22 364 91 11

◂ *Move over: king penguins at Volunteer Point on East Falkland Island must share their breeding colony with introduced domestic sheep.*

▲ *Reptile in retreat: feral goats are causing the rapid desertification of Alcedo Volcano on Isabella Island, the last stronghold of the giant Galápagos tortoise.*

Acknowledgements

Many people say the world is becoming noisier. In my experience, forests are becoming altogether quieter as more and more species disappear.

Growing up in New Zealand, I saw at first-hand the disastrous effect of invasive alien species on endemic bird populations. When I began to travel and work abroad, I realised that this was a recurring problem on most islands of the world.

In 2009, my cameraman husband Steve Hills and I made a film about invasive species in the UK Overseas Territories for the Royal Society for the Protection of Birds (RSPB). Seeing the devastation caused to seabird colonies in the South Atlantic prompted the writing of this book.

My thanks therefore go first and foremost to Steve, who not only makes inspiring films but has also suffered me reading out loud every word in this book. He knows at least as much about the subject as I do. I'm also very grateful to my editor Mike Unwin, publisher Donald Greig, project manager Anna Moores and the team at Bradt, as well as Chris Lane of Artinfusion for his excellent maps and design work. Thanks, too, to Megan Williams for her research into some of New Zealand's alien nasties.

I'd like to acknowledge the amazing work done on invasive species by the RSPB, who are protecting some of the most threatened wildlife habitats on Earth. Thanks, also, to Colin Clubbe and his colleagues at the Royal Botanic Gardens, Kew, and to Chela Zabin of the Smithsonian for kindly supplying the Caribbean barnacle picture.

Particular appreciation goes to the Global Invasive Species Programme and its exhaustive database that was the starting point for much of my research.

Most of all I'd like to salute the scientists, conservation workers, volunteers and other dedicated people around the world who are working so hard to combat this terrible threat to biodiversity.

Gill Williams, May 2011

Picture credits

Credits are listed in page order, specifying the position of the image on the page where necessary. For front and back cover credits, see page 2.

P4: Gary K Smith/FLPA; p6: Michael Durham/Minden Pictures/FLPA; p7: Michael Rose/FLPA; p8–9: Jurgen & Christine Sohns/FLPA; p10: Gerard Lacz/FLPA; p11: David Hosking/FLPA (main), Albert Lleal/Minden Pictures/FLPA (inset); p12: Thomas Marent/Minden Pictures/FLPA; p13: Heidi & Hans-Juergen Koch/Minden Pictures/FLPA; p14: Thomas Marent/Minden Pictures/FLPA; p15: Kathie Atkinson/OSF/Specialist stock (centre), Chris Mattison/FLPA (right); p16: Kevin Schafer/Minden Pictures/FLPA; p17: Wendy Dennis/FLPA; p18: Pete Oxford/Minden Pictures/FLPA; p19: Jurgen & Christine Sohns/FLPA (left), Pete Oxford/Minden Pictures/FLPA (right); p20: Scott Linstead/Minden Pictures/FLPA (centre), Ingo Schulz/Imagebroker/FLPA (inset right); p21: Michael & Patricia Fogden/Minden Pictures/FLPA; p22: Michael & Patricia Fogden/Minden Pictures/FLPA; p23: Michael & Patricia Fogden/Minden Pictures/FLPA; p24: Mark Moffett/Minden Pictures/FLPA (inset), Pete Oxford/Minden Pictures/FLPA (main); p25: Tui De Roy/Minden Pictures/FLPA; p26: Robin Chittenden/FLPA (left), Fabio Pupin/FLPA (right); p27: S & D & K Maslowski/FLPA (main), Imagebroker/FLPA (inset); p28–9: Do Van Dijck/Minden Pictures/FLPA; p30: Imagebroker/FLPA (main), Tui De Roy/Minden Pictures/FLPA (inset); p31: Hiroya Minakuchi/Minden Pictures/FLPA (main), Simon Litten/FLPA (inset); p32: Pat Morris/Ardea; p33: Paul Hobson/FLPA (both); p34: Aditya Singh/Imagebroker/FLPA; p35: Terry Whittaker/FLPA (top), Derek Middleton/FLPA (bottom); p36: Mitsuaki Iwago/Minden Pictures/FLPA (main), Wayne Hutchinson/FLPA (inset); p37: ImageBroker/FLPA; p38–9: S & D & K Maslowski/FLPA; p. 38: Derek Middleton/FLPA (top); p39: S & D & K Maslowski/FLPA (top); p40: Yva Momatiuk & John Eastcott/Minden Pictures/FLPA (left), Marcel van Kammen/Minden Pictures/FLPA (centre); p41: Pius Koller/Imagebroker/FLPA (left), Cyril Ruoso/Minden Pictures/FLPA (right); p42: Mike Lane/FLPA (main), Hugh Clark/FLPA (inset); p43: ImageBroker/FLPA (both); p44–5: S & D & K Maslowski/FLPA; p44: Jeremy Early/FLPA (left), Jules Cox/FLPA (right); p45: David Hosking/FLPA (top); p46: Derek Middleton /FLPA (left), Tui De Roy/Minden Pictures/FLPA (right); p47: Erica Olsen/FLPA (top), Brian Enting/Photo Researchers/FLPA (bottom); p48: Derek Middleton/FLPA (main), Jurgen & Christine Sohns/FLPA (inset); p49: Imagebroker/FLPA (left), Erica Olsen/FLPA (right); p50–1: Jurgen & Christine Sohns/FLPA; p50: Imagebroker/FLPA (top), Mitsuaki Iwago /FLPA (right); p52: Terry Whittaker/FLPA; p53: Gordon Roberts/FLPA (inset), Colin Monteath/FLPA (bottom); p54: Gerhard Zwerger-SC/Imagebroker/FLPA (top), Tui De Roy/FLPA (bottom left), Inga Spence/FLPA (bottom right); p55: Krystyna Szulecka/FLPA (background), Neil Bowman/FLPA (top); p56: Pete Oxford/FLPA; p57: Mitsuaki Iwago/FLPA; p58: Shaun Barnett/FLPA (left), Colin Monteath/FLPA (right); p59: Tui De Roy/FLPA (main), Mitsuaki Iwago/Minden Pictures/FLPA (inset); p60–1: Reinhard Dirscherl/FLPA; p62: ImageBroker/FLPA; p63: Chris Brignell/FLPA (top), Linda Lewis/FLPA (bottom); p64: Wil Meinderts/FN/Minden/FLPA (main), Foto Natura Stock/FLPA (inset); p65: Norbert Wu/FLPA (both); p66: Scott Leslie/Minden Pictures/FLPA (background), Ingo Arndt/Minden Pictures/FLPA (centre), John Eveson/FLPA (right); p67: Chela Zabin; p68: Malcolm Schuyl/FLPA (both); p69: Norbert Wu/Minden Pictures/FLPA; p70: Frank W Lane/FLPA (top), Terry Whittaker/FLPA (bottom); p71: Norbert Wu/FLPA (top), Fred Bavendam/FLPA (bottom and background); p72: R.Dirscherl/FLPA; p73: Reinhard Dirscherl/FLPA; p74: Michael Durham/FLPA (both); p75: Mark Newman/FLPA (top), Reinhard Dirscherl/FLPA (bottom); p76: Bob Gibbons/FLPA; p77: Michael Gore/FLPA (centre), Victoria Stone/OSF/Specialist Stock (right); p78: Imagebroker/FLPA (left), David Hosking/FLPA (right); p79: Yva Momatiuk & John Eastcott/FLPA (top), Science Source/Photo Researchers/FLPA (bottom); p80–1: Ingo Arndt/FLPA; p82: Gregory G. Dimijian/Photo Researchers/FLPA (top), Malcolm Schuyl/FLPA (bottom); p83: Nigel Cattlin/FLPA (main), Mark Moffett/FLPA (inset); p84: Mitsuhiko Imamori/FLPA (main), Peter Llewellyn/FLPA (inset); p85: USDA/Photo Researchers/FLPA (background and centre); p86: Barbara Strnadova/Photo Researchers/FLPA (centre), Ted Kinsman/Photo researchers/FLPA (bottom); p87: Gianpiero Ferrari/FLPA; p88: Treat Davidson/FLPA (both); p89: Nature's Best/Photo researchers/FLPA; p90: Konrad Wothe/Minden Pictures/FLPA (background), Nigel Cattlin/FLPA (inset); p91: Peter E. Smith/FLPA; p92: Mitsuhiko Imamori/FLPA; p93: Fritz Polking/FLPA (top), Shem Compion/FLPA (bottom); p94: Ingo Arndt/FLPA; p95: Scott Leslie/FLPA (top), Nigel Cattlin/FLPA (bottom); p96–7: Panda Photo/FLPA; p96: Bob Gibbons/FLPA (main), Frans Lanting/FLPA (bottom); p97: Norbert Wu/FLPA; p98–9: Imagebroker/FLPA; p100: Gary K Smith/FLPA (top), Jeremy Early/FLPA (bottom); p101: Mike Unwin; p102: Neil Bowman/FLPA; p103: Neil Bowman/FLPA (left), Hugh Lansdown/FLPA (centre); p104–5: Mike Lane/FLPA; p104: Mark Newman/FLPA (left), Imagebroker/FLPA (top); p105: Neil Bowman/FLPA (top); p106: S & D & K Maslowski/FLPA (main), Donald M. Jones/FLPA (inset); p107: Neil Bowman/FLPA; p108: Jeremy Early/FLPA (main), Philip Perry/FLPA (inset); p109: Klaus-Werner Friedri/Imagebroker/FLPA; p110: Herbert Kehrer/Imagebroker /FLPA (main), Roger Wilmshurst/FLPA (inset); p111: Roger Tidman/FLPA (top), Paul Hobson/FLPA (bottom); p112: Reinhard Dirscherl/FLPA (top), Erica Olsen/FLPA (bottom); p113: Simon Hosking/FLPA (top), Gerard Lacz/FLPA (bottom); p114–5: Colin Monteath/Minden Pictures/FLPA (background); p114: Kevin Schafer/Corbis Flirt/Alamy (left), Krystyna Szulecka/FLPA (centre); p115: Yva Momatiuk & John Eastcott/FLPA (top), Yva Momatiuk & John Eastcott/Minden Pictures/FLPA (right); p116: Imagebroker/FLPA (main), Simon Litten/FLPA (inset); p117: Donald M. Jones/FLPA (bottom), Terry Whittaker/FLPA (right); p118: Malcolm Schuyl/FLPA (main), Roger Wilmshurst/FLPA (inset); p119: Egon Bömsch/Imagebroker/FLPA; p120: Chris Brignell/FLPA (main), Philip Perry/FLPA (inset); p121: Martin B Withers/FLPA (main), Jurgen & Christine Sohns/FLPA (inset); p122: Mike Lane/FLPA (bottom left), Ingo Arndt/Minden Pictures/FLPA (top right); p123: Paul Hobson/FLPA (centre), Imagebroker/FLPA (right); p124–5: David Burton/FLPA; p126–7: Parameswaran Pillai Karunakaran/FLPA (background); p126: David Hosking/FLPA (top), Frans Lanting/FLPA (centre); p127: Inga Spence/FLPA (centre); p128: David Hosking/FLPA (background), Gary K Smith/FLPA (bottom left); p129: Nicholas and Sherry Lu Aldridge/FLPA (background), Nigel Cattlin/FLPA (top right); p130–1: Frans Lanting/FLPA (background); p130: Bettina Strenske/Imagebroker/FLPA (main); p131: Frans Lanting/FLPA (top and right); p132: Cyril Ruoso/FLPA (top), K. Wothe/Blickwinkel/Still Pictures (centre); p133: Leo Batten/FLPA (top), Nick Spurling/FLPA (bottom); p134: Richard Becker/FLPA (background), Michael Krabs Imagebroker/FLPA (centre); p135: Imagebroker/FLPA (top), Neil Bowman/FLPA (bottom); p136–7: Wayne Hutchinson/FLPA (background); p136: Roger Wilmshurst/FLPA (top), Tony Hamblin/FLPA (left); p137: John Adams CB MVO/FLPA (inset); p138: Inga Spence/FLPA (background), Daniel Roberts/FLPA (left); p139: Terry Whittaker/FLPA; p140: Leo Batten/FLPA (top), Nigel Cattlin/FLPA (centre); p141: Elliott Neep/FLPA (main), Nicholas and Sherry Lu Aldridge/FLPA (inset); p142: Nigel Cattlin/FLPA (background), Inga Spence/FLPA (main); p143: Tui De Roy/FLPA (main), Tui De Roy/FLPA (inset); p144–5: Suzi Eszterhas/FLPA (background); p144: Flip Nicklin/Minden Pictures/FLPA (centre); p145: NHPA/Franco Banfi (centre); p146: Klaus-Werner Friedri/FLPA; p147: Imagebroker/FLPA (top), Erica Olsen/FLPA (bottom); p148: Imagebroker/FLPA (background), Paul Hobson/FLPA (main); p149: Erica Olsen/FLPA (background), Bjorn Ullhagen/FLPA (inset); p150: Ingo Arndt/FLPA (background), Ray Goldring/Minden Pictures/FLPA (centre); p151: Chris and Tilde Stuart/FLPA (top), Chris Mattison/FLPA (bottom); p152: Nick Spurling/FLPA (top right), Tony Wharton/FLPA (bottom left), Terry Whittaker/FLPA (bottom right); p153: Michael Krabs/Imagebroker/FLPA; p154: Luciano Candisani/Minden Pictures/FLPA; p156–157: Tui De Roy/Minden Pictures/FLPA

Index

Main entries indicated in **bold**

albatross, black-browed 115
alligator, American 97
Antarctica 144–145
antelope, sable 140
ant, big-headed 95
little fire 95
red fire 80–81
yellow crazy 94

balsam, Himalayan 122, 148
bamboo 124–125
golden 134
Banks, Joseph 78–79, 150
barnacle, Caribbean 63, **67**
bass 67
largemouth 70
smallmouth 64
bats 31
bees, *Exoneura* 140
Killer 93
beetle, Asian longhorn 82, **86**
cochineal 139
coffee berry borer 142
French's 14
greyback cane 14
rhinoceros 102
bellbird 58
birds 100–101
bison 53
blackberry 147
blackbird 100
Bligh, William 78
bluebells 127
bluebird 104
bluegrass 144
boa, Mona Island 22
Puerto Rican 22
booby, Abbott's 94
breadfruit 78
bulbul, red-vented 107
bullfrog, American 12
bunting, Gough 48
butterflies, endangered 83
butterfly, Miami blue 19
orange monarch 107
white monarch 107
buttonquail, black-breasted 140

caiman, spectacled 10
caimans 10, 27, 96
camel 54
carp, common 75
cat, domestic 36–37, 96
caterpillar, cactoblastis 139
catfish, walking 76
cattle 30
chameleon, veiled 20
chicken 112–113
chough 122
cichlid 77, 97
Lake Victoria 63
coconut 143
coffee 142
coot 117
coral reefs 65, 71, **72–73**
cormorant, double-crested 96
coyote 117
coypu 31, 42
crab, land 94
north Atlantic spider 144
crayfish, American signal 68
white-clawed 68
crocodile, American 97, 151
Nile 77
crow, Indian house 102
crows 100

damsel, red 83
Darwin, Charles 6, 67, 78–79, 115, 119
deer, Chinese water 123
huemul 53
muntjac 123
red 52, **53**
reindeer 115
roe 53
sika 52
devil, Tasmanian 40
dingo 40
diver, red-throated 38
dodo 32
dog, feral 30
donkey 54
dormouse, edible 49
dove, Galápagos 119
duck, African black 116
American black 116
grey 116
Laysan 116
mandarin 118
muscovy 116
ruddy 118
white-headed 118
yellow-billed 116
ducks 101
dunlin 33

eagle, bald 117
ebony, St Helena 114
egrets 111–112
emus 100
eucalyptus 127, 150
Everglades 96–97

fig, Hottentot 122
finch, mangrove 24
finches, invasive 100
flax, New Zealand 78
flu, avian 113
flying fox 37
fox, red 28–29, 31, 40
San Joaquin kit 40
frigatebird, Ascension 115
frog, African clawed 13
Californian red-legged 13
Caribbean coqui 16
greenhouse 16
Pacific chorus 12
spotted tree 74
white-bellied 55
yellow-legged 74
frogs (endangered) 74

Galápagos Islands 24–25, 57, 156
gardening, for biodiversity 136–37
gecko, Madagascar day 11
pelagic 32
ginger, wild 135
goose, Canada 117
Egyptian 100, 116
giraffe, five-horned 123
glider, sugar 103
goat 24–25, **57**, 156
goby, tidewater 13
grass, bent 144
groupers 73
guillemots 101

harrier, hen 121
hedgehog, European 33
heron, great blue 76
hogweed, giant 149
honeycreeper, akepa 109
amakihi 109
palila 101
hummingbird, Costa's 100
hyacinth, water 127, **141**

ibis, African sacred 100, **111**
Australian 14
iguana, blue 8–9, **18**
green 18
insects 82–83

jellyfish, comb 66
junglefowl, grey 113
Himalayan 113
Sri Lankan 113

kangaroo, rat (bettong) 40
kelp, Japanese 63, 146
kingfisher, Mangaia 103
Pied 77
kipunji 31
kite, black 14
kiwis 33, 35, 59, 100
knotweed, Japanese 122

ladybird, harlequin 87
two-spot 87
lamprey, sea 69
lantana 103, 107, **140**

lapwings 33
leopard, clouded 31
lionfish 60–61, **71**
lizard, Tenerife spectacled 11
lizards 10–11
loon, common 110
lynx, Iberian 150

macaque, crab-eating 41
Magellan 79
mallard 116, 118
mammals 30–31
maple, Japanese 122
miconia 126
mink, American 31, **38–39**
European 38–39
mockingbird, Floreana 24
mongoose, Indian 16, **34**
monitor, Nile 10–11, **17**, 96
monkey-puzzle tree 122
moorhen 117
mosquito, *anopheles* 84
Asian tiger 84
moth, gypsy 82
mouse, house 48, 114
muskrat 39
mussel, zebra 64
myna, common 100, **103**, 140

noddy, brown 115, 120
lesser 120
numbat 31
nuthatch, European 108

oak, Californian 150
European 150
olive, St Helena 114
otter, Eurasian 39
owl, barn 120
barred 121
burrowing 9, 16
Eurasian eagle 121
northern spotted 121

panther, Florida 37, 96–97
parakeet, monk 108
ring-necked 7, **108**
parrot, Mauritius 103
Mauritius broad-billed 41
Seychelles black 108
superb 103
pear, prickly 24, 43, **139**
penguin, emperor 145
king 154
Magellanic 56, 79
penguins 101
perch, Nile 77
petrel, Jamaican 34
white-chinned 115
pigeon, feral 98–99, **119**
imperial 37
pink 34, 41
rock 119
pig, feral 31, **55**, 96
pine, Australian 151
plovers 33
pochard, red-crested 118
possum, brushtail 58–59
mountain pygmy 40
python, Burmese 22–23, 96–97

quolls 14, 40

rabbit, Amami 34
European 50–51
Lower Keys marsh 36
rail, bar-winged 34
Galápagos 25
rat, black 46, 115
Norway (brown) 6, **47**
Pacific 47
Sanibel rice 16
reed warbler, Cook Islands 103
reptiles 10–11
rhododendrons 122, 126, **152**
Royal Botanic Gardens, Kew
114–115
Royal Society for the Protection
of Birds (RSPB) 114-115

salamander, Californian tiger 12
sheep 56, 154
shoveler, Cape 116
shrew, Indian musk 32
silktail 107
white-toothed 35
skink, Whitaker's 35
slider, red-eared 26
snakehead, northern 70
snail, African land 88
cannibal 88
Carelia 88
drupella 73
O'ahu tree 88
Partula faba 88
snake, brown tree 20–21
Hispaniola 34
viperine 11
squirrel, barbary ground 43
grey 44–45
red 44–45
South Atlantic Islands 114–115
sparrow, house 100, **104–105**, 108
starfish, crown-of-thorns 65, 72
Starling, European 100, **106**, 108
stickleback, three-spine 13
stoat 4, **35**
stonechat, Canary Island 43
sunbird, Seychelles 120
swallow, cliff 104
tree 104
swan, black 111
mute 110
trumpeter 110

termite, Formosan 85
tern, Arctic 39
fairy 120
roseate 120
sandwich 111
sooty 115
terrapin, red-eared 26
thorn, Mexican 147
toad, cane 10, 14–15
Mallorcan midwife 11
Sulawesi 94
tobacco, tree 135
tortoise, desert 11, 54
giant 24–25, 37, 55, 95, 147, 156
trout, Arizona 74
cutthroat 74
gila 74
rainbow 74
tuatara 59
tui 78
turtle, common snapping 27
European pond 26
green 36–37, 114–115, 151
hawksbill 34, 36
loggerhead 43, 151
Pacific pond 26
red-bellied 26
triton 65

vole, water 39, 68
Victorians 122–123
vine, kudzu 126–127, 138

wallaby, Bennett's 123
brush-tailed rock 40
red-necked 123
wasp, common 82, 101
wattle, black 151
weasel 35
weed, white 114
weta 35, 59, 83
white-eye, Japanese 109
wirebird 103
woodpecker, green 100
wren, Stephens Island 37

zebra 123

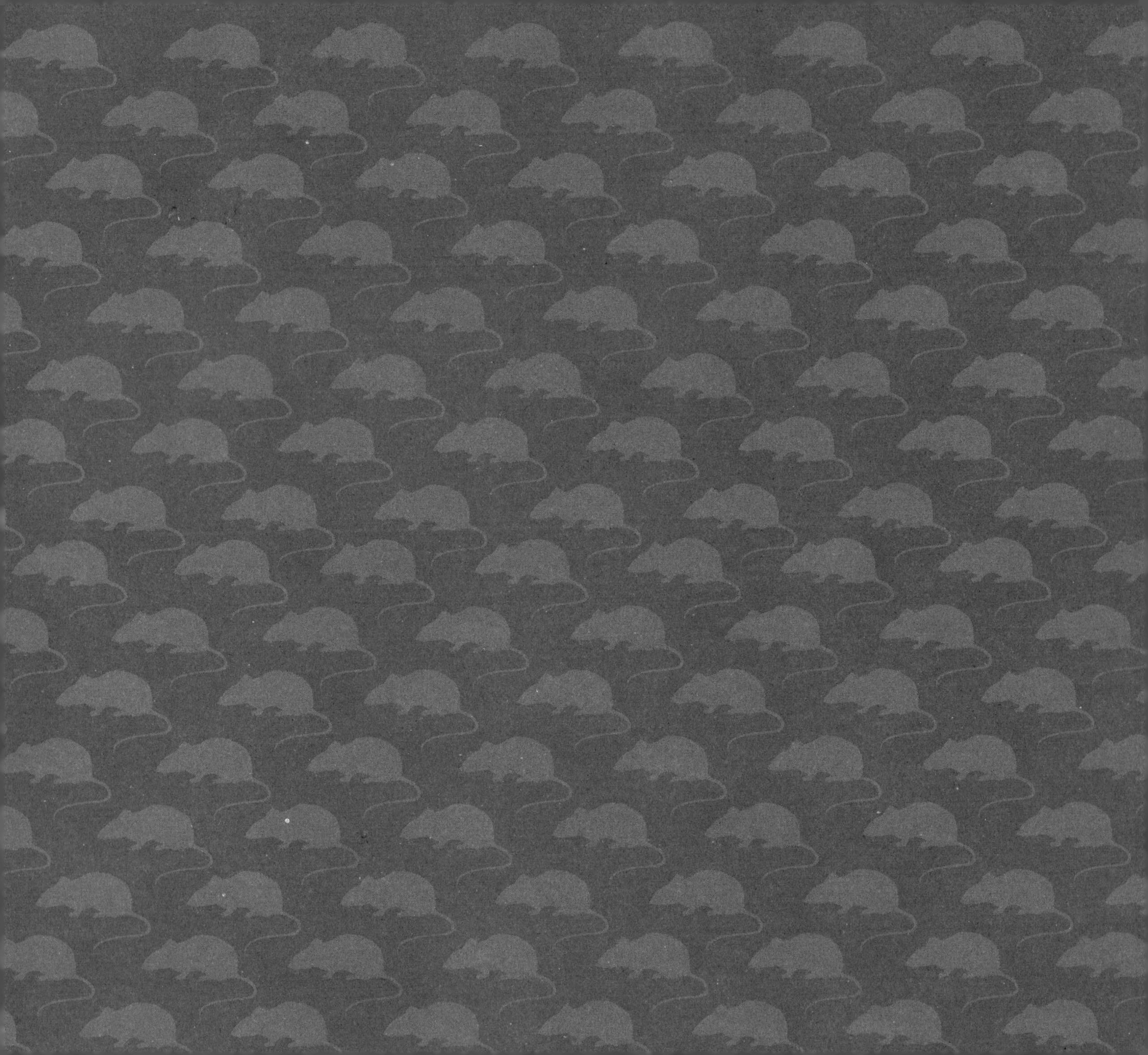